T0134601

Studies in Systems, Decision and Control

Volume 298

Series Editor

Janusz Kacprzyk, Systems Research Institute, Polish Academy of Sciences,
Warsaw, Poland

The series "Studies in Systems, Decision and Control" (SSDC) covers both new developments and advances, as well as the state of the art, in the various areas of broadly perceived systems, decision making and control–quickly, up to date and with a high quality. The intent is to cover the theory, applications, and perspectives on the state of the art and future developments relevant to systems, decision making, control, complex processes and related areas, as embedded in the fields of engineering, computer science, physics, economics, social and life sciences, as well as the paradigms and methodologies behind them. The series contains monographs, textbooks, lecture notes and edited volumes in systems, decision making and control spanning the areas of Cyber-Physical Systems, Autonomous Systems, Sensor Networks, Control Systems, Energy Systems, Automotive Systems, Biological Systems, Vehicular Networking and Connected Vehicles, Aerospace Systems, Automation, Manufacturing, Smart Grids, Nonlinear Systems, Power Systems, Robotics, Social Systems, Economic Systems and other. Of particular value to both the contributors and the readership are the short publication timeframe and the world-wide distribution and exposure which enable both a wide and rapid dissemination of research output.

** Indexing: The books of this series are submitted to ISI, SCOPUS, DBLP, Ulrichs, MathSciNet, Current Mathematical Publications, Mathematical Reviews, Zentralblatt Math: MetaPress and Springerlink.

More information about this series at http://www.springer.com/series/13304

Vitaliy Babak · Volodymyr Isaienko ·
Artur Zaporozhets
Editors

Systems, Decision
and Control in Energy I

 Springer

Editors
Vitaliy Babak
Institute of Engineering
Thermophysics of NAS of Ukraine
Kyiv, Ukraine

Volodymyr Isaienko
National Aviation University
Kyiv, Ukraine

Artur Zaporozhets
Department of Monitoring and Optimization
of Thermophysical Processes
Institute of Engineering Thermophysics
of NAS of Ukraine
Kyiv, Ukraine

ISSN 2198-4182   ISSN 2198-4190   (electronic)
Studies in Systems, Decision and Control
ISBN 978-3-030-48585-6   ISBN 978-3-030-48583-2   (eBook)
https://doi.org/10.1007/978-3-030-48583-2

This Springer imprint is published by the registered company Springer Nature Switzerland AG
The registered company address is: Gewerbestrasse 11, 6330 Cham, Switzerland

Preface

The concept of "energy" is meant here very generally and includes methods for obtaining and using various types of energy for the needs of human society. Energy is one of the foundations for the development of modern society. The effectiveness of solving social, economic and technical problems, as well as the anthropogenic transformations of nature, is largely determined by energy production and the scale of energy production.

Modern energy is not a separate industry, but it penetrates widely into other areas, in particular, chemical, transport, aerospace, construction, metallurgy, engineering, agriculture, etc. The energy sector is based on complex technical systems that are multicomponent, spatially distributed systems that during their operation are affected to a wide range of design and non-design thermomechanical loading conditions, the effects of aggressive fields and units, unauthorized influences (operator errors, terrorism, sabotage) and can reach various limit states.

Complex technical systems are characterized by complex non-linear interactions between their constituent elements, complex chains (scenarios) of cause–effect relationships between hazardous, probabilistic events and processes that occur during their life. These scenarios can be implemented over complex ramified scenario trees.

Ensuring the operational reliability, durability and safety of power equipment is a difficult task, which is associated with the organization of the reliability of control over the operation of power plants and ensuring optimal conditions for their operation. In this regard, we can distinguish a whole class of tasks related to the development of control systems, diagnostics and monitoring in the energy industry, which are presented in this book. Of particular relevance now is the use of UAVs in the energy sector.

Particular attention must be paid to the environmental consequences of the operation of energy facilities, the main of which is significant environmental pollution in large cities and industrial areas.

The development of environmental management information systems is the prerogative of the state, corporations and one of the main directions of the national informatization policy. A clearly debugged system of environmental monitoring gives a general idea of the features of the current ecological state, the main directions of state policy in the field of environmental protection and the use of natural resources and environmental safety. The methodology and hardware–software tools for monitoring the state of the environment presented in the monograph are effective tools for supporting decision-making in managing the environmental safety of the atmosphere during its technogenic pollution.

This book presents a comprehensive look at the current state and prospects for the development of energy and related industries, formed by a team of authors from various scientific institutions of Ukraine. Among the authors of the book are employees of: Institute of Engineering Thermophysics of NAS of Ukraine, Institute of Electrodynamics of the NAS of Ukraine, Pukhov Institute for Modelling in Energy Engineering of NAS of Ukraine, State Institution "Institute of Environmental Geochemistry" of NAS of Ukraine, Institute of General Energy of NAS of Ukraine, National Aviation University, M. E. Zhukovsky National Aerospace University "Kharkiv Aviation Institute", Ivano-Frankivsk National Technical University of Oil and Gas, Institute of Information Technologies and Learning Tools of NAES of Ukraine, Zhytomyr National Agroecological University, Zhytomyr Military Institute, National University of Life and Environmental Sciences of Ukraine, National University of Civil Defense of Ukraine, SE "state scientific and technical center for nuclear and radiation safety".

A special contribution to the creation of this book belongs to the Council of Young Scientists of the Department of Physical and Technical Problems of Energy of the National Academy of Sciences of Ukraine.

Kyiv, Ukraine   Vitaliy Babak
                                                                                      Volodymyr Isaienko
                                                                                       Artur Zaporozhets
                                                                                               Editors

Contents

About the Editors

Vitaliy Babak Corresponding Member of the National Academy of Sciences of Ukraine, Doctor of Technical Sciences, Professor, Honored Scientist of Ukraine, Laureate of the State Prize of Ukraine in the field of science and technology.

Affiliation: Institute of Engineering Thermophysics of NAS of Ukraine, Vice Head for Science

Address: 2a, Marii Kapnist (Zhelyabova) Str., Kyiv, 03057, Ukraine

e-mail: vdoe@ukr.net

Volodymyr Isaienko Academician of the National Academy of Higher Education Sciences of Ukraine, Doctor of Biological Sciences, Candidate of Technical Sciences, Professor.

Affiliation: National Aviation University, Rector

Address: 1, Ljubomyra Guzara Ave., Kyiv, 03058, Ukraine

e-mail: volodymyr.isaienko@gmail.com

Artur Zaporozhets Candidate of Technical Sciences, Senior Researcher, Laureate of the Presidential Award for Young Scientists.

Affiliation: Institute of Engineering Thermophysics of NAS of Ukraine, Senior Research Officer in Department of Monitoring and Optimization of Thermophysical Processes

Address: 2a, Marii Kapnist (Zhelyabova) Str., Kyiv, 03057, Ukraine

e-mail: a.o.zaporozhets@nas.gov.ua

Mathematical Approaches for Determining the Level of Impact of Ash-Slag Dumps of Energy Facilities on the Environment

Andrii Iatsyshyn(ID), Volodymyr Artemchuk(ID), Artur Zaporozhets(ID), Oleksandr Popov(ID), and Valeriia Kovach(ID)

Abstract The article is devoted to the problem of the impact of storage sites of ash-slag dumps of enterprises of the fuel and energy complex on the environment. The processes occurring in ash dumps are described, specifically: evaporation of water with the formation of dry ash and dust particles in the territory of dry sites under the influence of wind and water infiltration (illuminated or even partially untreated) and the penetration of dissolved forms of toxic ash-slag components into groundwater and water bodies located beyond them. The features of the transport of pollutants in the zone of influence of ash-slag dumps are investigated. The processes and phenomena that are observed during the transport of dissolved substances in the soil are determined. The components of the process of wind erosion and the mechanisms of action on particles located on the surface layer of the territory of ash dumps are described. The factors determining the erosion of ash-slag dumps as an areal object are listed. A mathematical model of the migration of pollutants in soils is described, which consists of a system of differential equations in partial derivatives of the second order with variable coefficients. An analytical solution of these equations is presented with certain simplifications for the case of two-dimensional flow, when the aquifer can be considered horizontal, single-layer and homogeneous. A two-dimensional model of the migration of soluble components is considered, taking into account the gradient of the relief as the main reason for the movement of water in the surface layers of the soil.

A. Iatsyshyn · V. Artemchuk · O. Popov
Pukhov Institute for Modelling in Energy Engineering of NAS of Ukraine, Kyiv, Ukraine

A. Iatsyshyn · V. Artemchuk · O. Popov · V. Kovach
State Institution "Institute of Environmental Geochemistry" of NAS of Ukraine, Kyiv, Ukraine

A. Zaporozhets (✉)
Institute of Engineering Thermophysics of NAS of Ukraine, Kyiv, Ukraine
e-mail: a.o.zaporozhets@nas.gov.ua

V. Kovach
National Aviation University, Kyiv, Ukraine

© The Editor(s) (if applicable) and The Author(s), under exclusive license to Springer Nature Switzerland AG 2020
V. Babak et al. (eds.), *Systems, Decision and Control in Energy I*, Studies in Systems, Decision and Control 298, https://doi.org/10.1007/978-3-030-48583-2_1

Keywords Ash-slag dump · Ash-slag waste · Migration model · Environmental pollution · Energy facilities · Thermal power plant · TPP

1   Introduction

Ash-slag dumps of enterprises of the fuel and energy complex (FEC) are special hydraulic storage facilities designed for storing ash-slag waste, the territory of which is limited by enclosing dams and terrain. In ash-slag dumps, two processes constantly occur. The first of these is the evaporation of water with the formation on the territory of the ash-slag dumps of the so-called "beaches"—dry areas with dust particles of ash and slag waste generated by the wind. The second process is the infiltration of water (illuminated or even partially untreated) and the ingestion of dissolved forms of toxic ash and slag components into groundwater and water bodies located outside ash-slag dumps. Harmful substances that are part of ash-slag dumps can migrate from the surface of the ash-slag dumps through air and water and pollute the surface layer of the atmosphere, soil, groundwater and surface water of areas located at a distance of several kilometers from the waste storage. Further spreading along different trophic chains, toxic substances contained in ash-slag waste can cause degradation of the biosystem (plant and animal life, aquatic organisms) and adversely affect human health (through water, air, food) [1].

Figure 1 shows the ash-slag dump of the Dobrotvorskaya thermal power plant (TPP). This TPP is located in Dobrotvor (Kamenka-Bugsky district) with a design capacity of 700 MW and is a structural unit of PJSC DTEK "Zakhidenergo". As indicated in [2], the source of heavy metals during the operation of this TPP is both fine particles of ash and ash entering the ash-slag dumps. The maximum amount of smoke emissions settles in the zone of 2.5–4.0 km from the TPP. Ash, blown out

Fig. 1 Ash-slag dump of Dobrotvor TPP

Fig. 2 "Dust storm" at the ash-slag dump of the Burshtyn TPP (June 2019)

of ash-slag dumps, is able to create several times more pollutant concentrations in the soil than as a result of emissions from pipes. It is the soil that is at the center of cross-border transport and migration of heavy metals. At present, up to 15 million tons of slag are stored on the territory of ash-slag dumps.

The "dust storms" can also occur in the territory of ash-slag dumps. An example of such an emergency situation at the ash-slag dump of the Burshtyn TPP is shown in Fig. 2.

Such situations are characteristic of most ash-slag dumps, especially in the summer months, when the water level in the pond-illuminators is greatly reduced or it dries almost completely when there are no operational methods of dust suppression. Capillary wetting of the surface of the layer, the use of irrigation as a way to combat dust, precipitation, melt water change the moisture content of the material and are those factors affecting the dusting of the surface of ash-slag dumps. Dusting is typical for areas where the moisture content of the layer surface is 1–3% (air-dry material), and is completely absent if humidity is more than 6–7%.

2 Literature Analysis and Problem Statement

A great contribution to the development of the theory of filtration is provided by the work of L.S. Basnieva, M. Musket, E. Scheidegger, V.L. Danilova, V.N. Nikolaevsky, M.D. Rosenberg, E. Settar, E.S. Romma, M.I. Shvidler, D.A. Efros and the like. The work of M.M. Verigina, A.V. Shibanova, B.S. Sherzhukova, V.I. Lavrik, A.P. Oleinik, V. M. Shestakova et al. about the transfer of pollutants in soils, groundwater and the surface air layer.

The environmental impact of the fuel and energy complex on the environment can be divided into the following categories:

- air pollution with harmful substances and compounds, taking into account sulfur dioxide, nitrogen oxides, solid particles and heavy metals, by their effect on human health, flora, fauna and the like;
- greenhouse gas emissions, taking into account carbon dioxide, methane, nitrous oxide, which contribute to global climate change;
- changes in the natural regime of water use and the negative impact on water quality through thermal and chemical pollution and the influence of hydroelectric power plants;
- change in the natural regime of land use by placing power plants and electric grids, removal and storage of waste, taking into account solid, liquid and nuclear waste.

The problem of the influence of the fuel and energy complex on the state of the environment is devoted to the scientific research of many scientists. In particular, in [3] the impact of the ash-slag dumps of the Tripol TPP on the health of the population living in adjacent territories was shown, the publication [1] examined the environmental impact of ash-slag dumps of TPPs, and in [4] the impact of some Chinese TPPs on environment is showed. Research [5] is devoted to the distribution of heavy metals in soils from emissions of a TPP, in [6] monitoring data are analyzed and the state of groundwater in the influence area of the Sumy TPP is estimated. The authors of [7] analyzed the harmful emissions of the Burshtyn TPP and the environmental situation around it. The development of equipment for measuring environmental quality parameters is described in [8, 9]. Some issues of constructing appropriate mathematical and software tools for assessing the impact of PECA on the economic component of the state and the natural environment are discussed in [10–18, 22].

However, it should be noted that in the cited works the issues of the transfer of pollutants in the zone of influence of the fuel and energy complex, especially in the areas of ash-slag dumps, the study of which is a necessary step in constructing the corresponding mathematical models of the process of migration of pollutants in the natural environment within the zones of ash-slag dumps and adjacent territories, are not sufficiently addressed.

This determines the need for research in this direction. Knowing the features of the mechanisms that determine the movement of pollutants in all components of an ecosystem, it is possible to create adequate mathematical models, which in turn will make it possible to obtain appropriate estimates of pollution levels in soils, groundwaters, and atmospheric air and the like.

3   Purpose and Objectives of the Study

The study was carried out with the aim of analyzing existing mathematical tools for assessing the impact of waste storage sites on the environment of the surrounding territories for their further development and the creation of appropriate software tools.

As part of the goal, it was proposed to solve the following tasks:

1. to determine the characteristics of the transfer of pollutants in the zone of influence of the fuel and energy complex.
2. to analyze a number of mathematical models of the migration of pollutants in soils, water, and in the surface air layer in the areas affected by ash-slag dumps.

4   Research Methods

This study was carried out mainly on the analysis of the publications of leading Ukrainian scientists on mathematical tools for assessing the impact of waste storage sites on the environment of adjacent territories.

The following methods were used in this study: comparative analysis method; analysis of the experience of ash and slag waste, methods for monitoring, modeling and assessing the impact of ash-slag waste storage sites on the environment and population.

5   Research Results

5.1   Features of the Transfer of Pollutants in the Zone of Influence of the Fuel and Energy Complex

The transfer of pollutants to groundwater occurs from the aeration zone of the fuel and energy complex, the industrial site and the ash dump site. Therefore, the study of the migration of pollutants differs during describing the process in the aeration zone and in the zone of influence of ash-slag dumps.

In the aeration zone of TPPs, there is an infiltration of pollutants, which, settling from the exhaust gases to the soil surface, are transported down the profile with precipitation and melt water. The rate of transport of pollutants will depend on the amount of precipitation, affect the saturation of the soil, and the ambient temperature, which will affect the evaporation of moisture from the surface layers of the soil. From the body of ash-slag dumps, the transfer of pollutants will be carried out in water-saturated soil due to filtration. During the transfer of dissolved substances in the soil, a number of features were revealed [6]:

- there is no clear border between the solution that enters the soil and soil moisture, so the front of the pollutant solution is "washed out";
- there is continuous mixing of the solution and soil moisture, resulting in the formation of an expanded dispersion zone;
- the process of mixing or "erosion" of the front of the solution of polluted effluents is the stronger, the higher the flow rate and large pores of the soil.

The transport of solutes in the soil is affected by: diffusion, convection, dispersion, physico-chemical and biological processes, such as ion exchange, radioactive decay, sorption, decomposition by bacteria and the like.

In dry periods of the year, most ash-slag dumps can be a source of intensive pollination as a result of wind exposure. Wind erosion of ash-slag fields is determined by the nature and intensity of the impact of the wind flow, the properties of the stored material, storage technology, as well as the design of the object. Three components of the process of wind erosion are distinguished, specifically: separation and take-off of particles from the surface; moving particles in a dusty stream above the ash-slag field surface and scattering of ash eroded particles outside the ash-slag dumps after the dust cloud leaves the dam [18].

The influence of the air flow on each individual particle on the layer surface in the territory of ash-slag dumps is associated with several simultaneously operating mechanisms. This frontal aerodynamic pressure, causing a shift in the direction of the wind along the surface, the difference in static pressure that occurs when the particle flows around and creates lift, and turbulent diffusion in the wind flow creates variables pulsating in magnitude and direction of impact on the particle and weakens gravitational and adhesive the connection of the particle with the surface of the layer.

Three possible pollination zones can be identified on the territory of ash-slag dumps: dry beaches on ash-slag fields; dams composed of ash-slag (non-sediment) dust deposits in the aerodynamic shadow of the dam (secondary pollination).

The erosion of the surface of a layer of dust particles depends on the conditions of the formation of the layer and the non-fuel properties of the particles. The most important surface parameters are exposed to wind:

- granulometric composition of the stored material;
- humidity of the surface layer of dust particles—technological (due to wetting by irrigation, residual moisture during fluctuation of the water level in ash-slag dumps, moistening of the surface layer with capillary moisture from sub-layer water volume) and as a result of precipitation;
- the density of the surface layer is determined by the method of washing ash (underwater or surface, dispersed or concentrated), using mechanical means of compaction of the surface layer;
- snow cover;
- vegetation cover on waste ash and slag dumps.

It is also worth noting that the erosion of ash dumps is determined by the following factors:

- protection of the object from wind exposure by the relief of the adjacent territory;

- the structure of the object, that is, the presence and height of the enclosing structures (dams, protective walls, forest planting around the perimeter of the ash dump);
- operational characteristics—the level of the layer surface along the dam, the surrounding territory, the area of dry beaches.

5.2 Mathematical Models of Migration of Pollutants in Soils

The mathematical model of migration [6], that is, the mass transfer of dissolved substances in the filtration flows of non-conservative pollutants, describes the interaction between soils and waste water, filtered using the equations of material balance and the kinetics equation. It consists of a system of second-order partial differential equations with variable coefficients, which, in the case of a three-dimensional plane-vertical (profile) constant filtration, provided that the convective diffusion coefficient is constant, has the following form (1–3):

$$\frac{\partial v_x}{\partial x} + \frac{\partial v_y}{\partial y} + \frac{\partial v_z}{\partial z} = 0; \quad v_x = \frac{\partial \varphi}{\partial x}; \quad v_y = \frac{\partial \varphi}{\partial y}; \quad v_z = \frac{\partial \varphi}{\partial z}, \tag{1}$$

$$D\left(\frac{\partial^2 c}{\partial x^2} + \frac{\partial^2 c}{\partial y^2} + \frac{\partial^2 c}{\partial z^2}\right) - v_x \frac{\partial c}{\partial x} - v_y \frac{\partial c}{\partial y} - v_z \frac{\partial c}{\partial z} - \frac{\partial N}{\partial t} = \sigma \frac{\partial c}{\partial t}, \tag{2}$$

$$\frac{\partial N}{\partial t} = \alpha(\sigma c - \beta N), \tag{3}$$

where $v = \{v_x, v_y, v_z\}$ is the filtration rate vector, m/day; $\varphi(x, y, z, t)$ is the filtration potential, m^3/day; $c(x, y, z, t)$ and $N(x, y, z, t)$ are the concentration of the substance diffusing, respectively, in the liquid and solid phases, kg/m^3; D is the convective diffusion coefficient, m^2/day; σ is the porosity of the medium, %; t is the time, day; α is the mass transfer coefficient, s^{-1}; β is the distribution coefficient of the substance between the phases under equilibrium conditions along the linear Henry isotherm, expressed by the equality $c_p = \beta N$.

To simplify the system of equations by averaging those substances that are sought (concentration, filtration rate, etc.), it is necessary to reduce the dimension of the system with one or more spatial coordinates. Vertical averaging over the z coordinate simplifies the equations in the two-dimensional profile model, which makes it possible to estimate the distribution of wastewater deep into the soil layer.

In the case of vertical filtering, Eqs. (4–6) are written in the form [19]:

$$\frac{\partial v_x}{\partial x} + \frac{\partial v_y}{\partial y} = 0; \quad v_x = \frac{\partial \varphi}{\partial x}; \quad v_y = \frac{\partial \varphi}{\partial y}; \quad \varphi = -\chi h; \quad h = \frac{p}{\rho g} - y, \tag{4}$$

$$D\left(\frac{\partial^2 c}{\partial x^2} + \frac{\partial^2 c}{\partial y^2}\right) - v_x \frac{\partial c}{\partial x} - v_y \frac{\partial c}{\partial y} - \frac{\partial N}{\partial t} = \sigma \frac{\partial c}{\partial t}, \tag{5}$$

$$c(x, y, t_0) = c_0(x, y); \quad c(x, y, t)|_{x=x_1} = c_0(y, t); \quad \frac{\partial c}{\partial z}\bigg|_{z=L} = 0, \tag{6}$$

where χ is the filtration coefficient, m/day; h is the pressure (O_y axis is directed vertically downwards), m; p is the pressure, Pa; ρ is the density, kg/m³; g is the acceleration of gravity, m²/s.

An analytical solution was shown in [20] in the case of a two-dimensional flow, when the aquifer is horizontal, single-layer, and homogeneous. The water velocity is considered constant and parallel to the O_x axis, and the diffusion coefficients are constant and proportional to the velocity. A pollutant is launched with a concentration of c_0 and a flow rate Q in the source. For the period dt, the mass of the pollutant launched will be $c_0 Q dt$. The initial concentration is zero. The area is not limited, and the concentration at infinity is zero. The equation describing this process has the form:

$$\alpha_L u \frac{\partial^2 c}{\partial x^2} + \alpha_T u \frac{\partial^2 c}{\partial y^2} - u \frac{\partial c}{\partial x} = \frac{\partial c}{\partial t}, \tag{7}$$

where α_L and α_T are the structural parameters depending on the type of soil, permeability and consolidation.

For the continuous flow of wastewater at flow rate Q, the solution of this equation for time t will have the form:

$$c(x, y, t) = \frac{c_0 Q}{2\pi u \sqrt{\alpha_L \alpha_T}} \int_0^t \exp\left[\frac{(x - u\theta)^2}{4\alpha_L u\theta} - \frac{y^2}{4\alpha_T u\theta}\right] \frac{d\theta}{\theta}. \tag{8}$$

In [21], a two-dimensional model of the migration of soluble components was proposed, taking into account the gradient of the relief as the main cause of water movement in the surface soil layers of an irrigated or periodically leaching type:

$$C = C_0 + \frac{C_p - C_0}{Be}\left[1 - \exp(-Bkt)\right]\exp(-\frac{k}{e_1}\tau), \tag{9}$$

where C is the average concentration of pollution in the soil solution of the infiltration layer, mg/m³; C_0, C_p are initial pollution concentrations in accordance with the infiltration flow and stagnant zones of the soil, mg/m³; $B = 1/e + 1/e_1$ is the structural parameter of the soil; e, e_1 are relative volumes of flowing and stagnant zones; k is the mass transfer coefficient, m/s; t is the integral advancement of the concentration front of pollution from the watershed to the point, s; τ is the integral time of precipitation for the control period, s.

5.3 Estimation of Emissions of Ash Particles From Ash-Slag Dumps into the Atmosphere Outside the Sanitary Protection Zone

The emission of ash particles into the atmosphere outside the sanitary protection zone of ash-slag dumps is determined by the formula [18], t/year:

$$M_B = M_E \cdot \left(\frac{1 - [\mu_0 - \mu_{DS}]}{\mu_0} \right), \tag{10}$$

where M_E is the annual removal of ash particles in each direction of the wind; μ_0 is the initial concentration of dust particles in the east from the dam; μ_{DS} is the dust content at the external border of the sanitary protection zone, determined by instrumental measurements or calculated by the formula.

The intensity of surface wind erosion depends on the method of layer formation. For the washed layer of ash-slag dumps, it is an order of magnitude lower than for the bulk and uncompressed layer. This indicates a significant difference in the pollination conditions of dry freshly washed areas and secondary pollination zones, the surface of which is formed by deposited eroded particles.

The transfer of ash particles entering the atmosphere from the open surface of ash-slag dumps and the underlying surface in the adjacent zone is carried out in accordance with a different mechanism than the dispersion processes in the atmosphere of solid particles with flue gases from TPPs and solid-state precipitation from a smoke plume.

The ground concentration of dust particles at a distance X from the dam taking into account background pollution is determined by the formula:

$$\mu_X^{BAC} = \mu^{BAC} + \mu_0 \times e^{-\alpha \times X} \tag{11}$$

where μ^{BAC} is the background dust content, mg/m^3 ≈ 0.3; X is the distance from the dam along the pollination axis, m (10, 50, 100, 250, 500, 1000).

The initial concentration of dust particles in the east from the dam (mg/m^3) is determined by the formula:

$$\mu_0 = \frac{M_{EP}}{L_\Pi \cdot h_0 \cdot U_{EF}}, \tag{12}$$

where L_Π is the flight of a saltos particle over an ash and slag field (width of a dust cloud), m; h_0 is the height of the dust cloud on the dam, m; and is found by the formula:

$$h_0 = 2 \cdot h, \tag{13}$$

where h is the elevation height of eroded particles (not counting the possible influence of ascending air currents and turbulent large-scale vortex formations).

The site immediately in front of the enclosing dam, the dam itself, its lower slope, drainage and upland ditches within the dump and the sanitary protection zone with natural and artificial dust-shelterbelts are a dust-free zone of the wind flow, in which gravitational forces and turbulent diffusion act.

The speed of the air flow determines not only the intensity of blowing ash particles from the surface of the ash-slag dumps (specific swelling per surface unit), but also the length of the pollination front (as a result of the involvement of new pollination sites dispersed over the surface of the dump) and the height of the dust cloud formed over ash-slag field.

During the wind speed at the level of the wind vane is close to U_{CR} (wind speed at which blown ash particles begin) pollination is local in nature. With an increase in wind speed, the mediated character of pollination is maintained; narrow plumes of dusty wind-air flow are transported through the dam. At a wind speed, which reaches 6.8 m/s, which corresponds to a maximum wind speed U_{max} with a repeatability of 5%, the pollination front (dust cloud width L_{Π}) increases to 0.5 of the length of dam L_{Π}. Approximately the width of the dust cloud can be determined from formula [18]:

$$L_{\Pi} = 0,5 \cdot L_D \frac{(U_Z - U_{CR})}{(U_{max} - U_{CR})} \tag{14}$$

where L_D is the length of the dam, m.

In the same range of wind speeds, the upper boundary of the dust cloud, moves above the surface of the ash-slag field, varies from 0.2 to 3 m.

At $U_Z > U_{CR}$, pollination can cover the surface of all potentially dusty areas of a dry "beach" (up to 20% of the area of a dry "beach"), and dust clouds will melt along the entire length of the dam, excluding the area shielded by a settling pond.

5.4 Particulate Emissions Swell From the Surface of Ash-Slag Dumps

The calculation of emissions of solid particles that swell from the surface of ash-slag dumps proposed to calculate according to the following formula [18], g/s:

$$M_{SP} = 86.4 \cdot K_0 \cdot K_1 \cdot K_2 \cdot S \cdot W \cdot j \cdot (365 - T) \tag{15}$$

where K_0 is a coefficient taking into account the moisture content of the material (0.1–1.0); K_1 is a coefficient taking into account wind speed (2–0.5); K_2 is a coefficient taking into account the efficiency of blowing solid particles (0.1–1.0); S is the surface area on which pollination occurs, m^2; W is the specific swellability of particulate matter from a dusty surface kg/m^2; j is a coefficient of grinding of rock mass (0.1–1.0); T is the annual number of days with stable snow cover.

6 Discussion of the Results

Mathematical models of the migration of pollutants in soils should be formed taking into account: the type of wastewater, the properties of the impurities that are in them and can interact with soil particles; soil characteristics, that is, rocks, structures, pore shapes and soil cracks, as well as the presence of moisture in its composition; processes of interaction between them. But during solving systems of equations, a number of difficulties arise related to their complexity.

Therefore, to reduce the dimension of model (1–3), it was proposed to average the sought quantities (concentration, velocity, etc.) over one or two spatial coordinates, which allowed us to obtain the analytical dependence (8). Such simplifications reduce the accuracy of models and increase computational errors.

The transfer of ash particles entering the atmosphere from the open surface of the ash dump and the underlying surface in the adjacent zone is carried out in accordance with a different mechanism than the processes of dispersion in the atmosphere from point stationary sources of pollution of the fuel and energy complex. In the absence of special equipment, the formulas proposed in the publication (10, 14, 15), based on the physical model of the wind erosion process, allow us to evaluate the effect of ash and slag dumps on the surface air layer.

7 Conclusions

The study and prediction of the state of individual elements of the natural environment under the conditions of anthropogenic load is one of the most important tasks of modern environmental research.

As a result of the studies, the features of the transport of pollutants in the zone of influence of enterprises of the fuel and energy complex, which provides the possibility of mathematical modeling of the migration of pollutants in soils and in the surface layer of air from ash and slag dumps, were established.

A number of mathematical models have been proposed and analyzed to study the process of migration of pollutants within the zones of ash and slag dumps and adjacent territories.

The above funds may be useful in the development of measures to reduce environmental pollution by enterprises of the fuel and energy complex. They can be recommended to design organizations when developing forecast estimates of the impact of new ash and slag dumps, as well as to regulatory authorities to assess pollution and improve the environmental performance of energy facilities.

References

1. Iatsyshyn, A., Popov, O., Artemchuk, V., Kovach, V., Kameneva, I.: The peculiarities of the thermal power engineering enterprise's ash dumps influence on the environment. Prob. Emerg. Situations **28**, 57–68 (2018). https://doi.org/10.5281/zenodo.2594489
2. Kovalchuk, O.P., Snitynskyy, V.V., Shkumbatyuk, R.S.: Monitoring of heavy metals content in soils of the areas surrounding Dobrotvir thermal power plant. Sci. Bull. UNFU **27**(4), 87–90 (2017). https://doi.org/10.15421/40270419
3. Report. Investigation of the hygienic estimation of the economic outflow of the Trypylskaya TPP, which is part of the Centralenergo PJSC, for pollution of the natural environment and living conditions of the population of the Obukhiv district of the Kyiv region, Kyiv, 2016
4. Jia, H., Zheng, S., Xie, J., Ying, X. Ming, Zhang, C.: Influence of geographic setting on thermal discharge from coastal power plants. Mar. Pollut. Bull. **111**(1–2), 106–114 (2016). https://doi.org/10.1016/j.marpolbul.2016.07.024
5. Levchenko, A.E, Ignatenko, M.I, Hobotova, E.B.: The heavy metal pollution of soil near thermal power plants. Energy Ecol. Hum, 462–468 (2013)
6. Miakaieva, H.M.: Modeling of technogenic influence of thermal power plants on the hydrosphere. Dissertation for a Candidate Degree of Engineering Sciences. Sumy State University, Ukraine (2018)
7. Kryzhanivsky, Ye.I., Kolhlak, G.V.: Ecological problems of energy. Naftogazovaya energetika **25**(1), 80–90 (2016)
8. Isermann, R.: Fault-diagnosis applications: model-based condition monitoring: actuators, drives, machinery, plants, sensors, and fault-tolerant systems. Springer-Verlag Berlin Heidelberg (2011). https://doi.org/10.1007/978-3-642-12767-0
9. Babak, V.P., Babak, S.V., Myslovych, M.V., Zaporozhets, A.O., Zvaritch, V.M.: Principles of construction of systems for diagnosing the energy equipment. In: Diagnostic Systems For Energy Equipments. Studies in Systems, Decision and Control, vol. 281, pp. 1–22. Springer, Cham (2020). https://doi.org/10.1007/978-3-030-44443-3_1
10. Popov, O., Iatsyshyn, A., Kovach, V., Artemchuk, V., Taraduda, D., Sobyna, V., Sokolov, D., Dement, M., Yatsyshyn, T.: Conceptual approaches for development of informational and analytical expert system for assessing the NPP impact on the environment. Nucl. Radiat. Saf. **79**(3), 56–65 (2018). https://doi.org/10.32918/nrs.2018.3(79).09
11. Popov, O., Iatsyshyn, A., Kovach, V., Artemchuk, V., Taraduda, D., Sobyna, V., Sokolov, D., Dement, M., Hurkovskyi, V., Nikolaiev, K., Yatsyshyn, T., Dimitriieva, D.: Physical features of pollutants spread in the air during the emergency at NPPs. Nucl. Radiat. Saf. **84**(4), 88–98 (2019). https://doi.org/10.32918/nrs.2019.4(84).11
12. Popov, O., Iatsyshyn, A., Kovach, V., Artemchuk, V., Taraduda, D., Sobyna, V., Sokolov, D., Dement, M., Yatsyshyn, T., Matvieieva, I.: Analysis of possible causes of NPP emergencies to minimize risk of their occurrence. Nucl. Radiat. Saf. **81**(1), 75–80 (2019). https://doi.org/10.32918/nrs.2019.1(81).13
13. Ilse, K., Micheli, L., Figgis, B.W., Lange, K., Daßler, D., Hanifi, H., Wolfertstetter, F., Naumann, V., Hagendorf, C., Gottschalg, R., Bagdahn, J.: Techno-economic assessment of soiling losses and mitigation strategies for solar power generation. Joule **3**(10), 2303–2321 (2019). https://doi.org/10.1016/j.joule.2019.08.019
14. Jones, R.K.: Solving the soiling problem for solar power systems. Joule **3**(10), 2298–2300 (2019). https://doi.org/10.1016/j.joule.2019.09.011
15. Stohniy, O.V., Kaplin, M.I., Bilan, T.R.: Methods and means of the account for the factors of energy safety in an economico-mathematical model of country's supply with fuel. Probl. Zagal'n. Energ. **31**(4), 38–45 (2012)
16. Stohniy, O.V., Kaplin, M.I., Bilan, T.R.: An economic mathematical model for the import of coal to Ukraine. Probl. Zagal'n. Energ. **28**(1), 29–34 (2012)
17. Stohniy, O.V., Kaplin, M.I., Bilan, T.R.: Features of a model of the production type for calculating the volumes of prospective fuel substitution in the fuel supply system of state's economy. Probl. Zagal'n. Energ. **30**(3), 30–36 (2012)

18. Komonov, S.V., Komonova, E.N.: Wind erosion and dust suppression. Lecture course. Krasnoyarsk: Publishing House SFU, Russia (2008)
19. Bojko, T., Abramova, A., Zaporozhets, J.: Mathematical modeling of polluting migration in soils. East-Eur. J. Enterp. Technol. **6**(4)(66), 14–16
20. Fried, J.J.: Groundwater pollution. Moscow (1975)
21. Shandiba, O.B., Varlamov, M.K., Martynenko, O.P., Borosenets, N.S.: Environmental monitoring of chemical migration in contaminated territories. SSAU Bulletin, series "Mechanization and automation of technological processes", vol. 5, pp. 69–71 (2000)
22. Babak, S., Babak, V., Zaporozhets, A., Sverdlova, A.: Method of statistical spline functions for solving problems of data approximation and prediction of objects state. In: CEUR Workshop Proceedings, vol. 2353, pp. 810–821. Access mode: http://ceur-ws.org/Vol-2353/paper64.pdf

Overview of Quadrocopters for Energy and Ecological Monitoring

Artur Zaporozhets⊙

Abstract The article provides an overview of serial small-sized quadrocopters that can be used for monitoring power plants, electric networks, pipelines, agriculture, land, forest and water resources, infrastructure, etc. A review of the technical characteristics and flight qualities of 8 types of quadrocopters. Their distinguishing feature is the presence of a special suspension, the ability to lift loads from 250 grams and cost up to $ 1000. In this case, the suspension can be used to mount measuring equipment or other sensors. An analysis of the presented quadrocopters based on the method of universal qualitative efficiency criterion was performed. The following characteristics were selected as parameters for analysis: price/quality, camera, functionality, flight time, flight qualities, compactness, equipment, reliability. Based on the analysis, the best types of quadrocopters that can be used to solve monitoring problems were selected. A prototype system for monitoring the technical condition of the main pipelines of heating networks is presented. This complex is based on one of the quadrocopters presented in the review.

Keywords Quadrocopters · UAVs · Drones · Monitoring · Technical characteristics · Payload · Analysis

1 Introduction

The increasing negative impact of man on all components of the environment requires the use of new, more effective means and methods of control. At the same time, one-time observations do not provide an idea of either the intensity of the ongoing processes, nor of their dynamics, and sometimes of the fact of the negative impact itself. That is why recently the implementation of monitoring systems in assessing the state of natural and man-made objects has become increasingly relevant [1, 2]. In this case, monitoring is understood as a complex system of observations of the state of the environment (atmosphere, hydrosphere, soil and vegetation cover, etc.) and

A. Zaporozhets (✉)
Institute of Engineering Thermophysics of NAS of Ukraine, Kyiv, Ukraine
e-mail: a.o.zaporozhets@nas.gov.ua

© The Editor(s) (if applicable) and The Author(s), under exclusive license
to Springer Nature Switzerland AG 2020
V. Babak et al. (eds.), *Systems, Decision and Control in Energy I*, Studies in Systems,
Decision and Control 298, https://doi.org/10.1007/978-3-030-48583-2_2

technogenic objects for the purpose of their control and protection [3]. At the same time, high efficiency and effectiveness of control can be achieved through the use of remote research methods performed from the quadrocopter [4]. Information received in a timely manner as a result of systematic monitoring allows making informed management decisions in the field of exploitation of natural and technogenic objects [5].

In this paper, an overview of serial quadrocopters with a cost of up to $ 1000 will be presented, the main task of which is photo and video shooting. Their distinctive feature is the presence of a special suspension for the camera, which can be used to mount various devices and sensors, as well as the ability to carry a payload of 250 grams [6].

The purpose of the work is the analysis of existing serial quadrocopters, which can be used for various monitoring in the fields of energy and ecology; comparison of their properties based on technical specifications declared by the manufacturer, as well as public opinion of their users [7, 8].

2 Quadrocopter Monitoring Tasks

Drones can effectively increase the security level of the perimeter of the power plant and individual infrastructure [9]. Regular monitoring using drones allows to monitor work on the territory of power plants, remotely carry out reconnaissance of the terrain and adjacent territories, and identify the presence of unauthorized persons and vehicles. A quadrocopter equipped with a gamma radiation detection unit allows reconnaissance of the radiation situation in the area [10]. Intelligence data can be transmitted from the quadrocopter to the ground control station in real time.

Monitoring high-voltage power lines using drones allows to quickly and efficiently assess the technical condition of objects in hard-to-reach and remote places [11].

Among quadrocopters tasks also are assessment of the condition of poles and insulators, level of sagging wires and the height of vegetation surrounding power lines, earlier identification of threats and acts of activity of unauthorized persons in protected areas [12].

Thermal imaging allows you to identify areas of thermal energy leakage and poor insulation [13, 14].

The data obtained with the help of quadrocopters allow specialists to assess risks and predict the impact of natural factors, to explore new routes for laying power lines and the territories adjacent to them [15, 16]. Approaches for diagnosing different energy equipment are given in [17–19]

Also an urgent task is to monitor air pollution using quadrocopters, which makes it possible to evaluate air quality at different heights, to predict the movement of concentration fields of pollutants [20, 21].

3 Quadrocopter Analysis

MJX X101. Model X101 (Fig. 1) refers to an inexpensive quadrocopter and is adapted for the installation of various types of equipment, such as cameras or video cameras. The features of this quadrocopter include:

- axle-free suspension with vibration dampers, adapted to set a small load;
- the function of returning the drone to the control panel;
- LED lights;
- 2 modes of expenses;
- ready to fly out of the box.

This quadrocopter is equipped with commutator engines, the potential of which is quite enough for a flight with a compact cargo without a significant loss in controllability. Torque transmission is implemented through the gear, which generally increases the noise of the quadrocopter.

The developer can bundle the MXJ X101 with 2 types of cameras- C4016/C4018. The shooting quality of both is below average. The camera is capable of broadcasting a video stream via Wi-Fi channel to smartphone (IOS/Android).

The quadcopter is also equipped with a powerful 2S LiPo battery for this class with a capacity of 1200 mAh with a discharge current of 25C. Such a battery is

Fig. 1 MXJ X101

capable of providing no more than 10 min of flight of a quadrocopter with maximum load. Battery charge time—2–2.5 h.

The quadcopter also includes a 4-channel control panel GR-246R, operating at a frequency of 2.4 GHz. Such a remote control provides the flight range of a quadrocopter to a distance of not more than 100 m. In a metropolis, this distance is not more than 50 m. The remote control is powered by 3 AA batteries.

During flight tests, the MJX X101 quadrocopter shows good dynamics and wind resistance with excellent vertical thrust. The function of returning the drone to the equipment (Headless) is also carried out without serious technical comments, however, it requires the involvement of an operator. In practice, the flight time of the MJX X101 does not exceed 10 min. Operation of the quadcopter along with the C4016/C4018 Wi-Fi camera significantly reduces its flight time (no more than 7 min).

In different stores, the cost of the MJX X101, depending on the configuration, is $75–120.

Syma X8HG. The quadcopter Syma X8HG (Fig. 2) is a continuation of the eighth series of drones from Syma. Externally, the model is almost no different from its "predecessors", but contains a number of minor and important technical advantages.

The features of this quadrocopter include:

- height maintenance function;
- Headless mode;
- control at a frequency of 2.4 GHz;
- 8 MP HD-camera;

Fig. 2 Syma X8HG

- 6 axis gyroscope;
- 360° flip;
- LED backlight.

One of the main differences of this drone from previous models is the presence of a height barometer, which will greatly facilitate the height maintenance function for a new user.

The Syma X8HG quadcopter is equipped with improved engines of the collector type FK-132VH-38–29/38, which are lighter than the motors installed on the X8C and X8W models.

The basic configuration of the drone is equipped with an 8 megapixel photo camera with a GoPro format video camera that shoots in 2 modes: 720p/1080p. The quality of the camera is average, and is suitable only for the first acquaintance with the device.

The quadcopter Syma X8HG is equipped with a 2S LiPo battery with a capacity of 2000 mAh with a discharge current of 25C. The flight time according to the specification is 5–7 min. In practice, it is possible to achieve a flight lasting 10 min. It takes about 70 min to fully charge the battery.

Complete with Syma X8HG is a standard for the 8th series of drones 4-channel equipment operating at a frequency of 2.4 GHz and provides a drone removal distance of up to 200 m in direct visibility. The equipment is powered by 4 AA batteries.

During flight tests, the Syma X8HG showed a difference in dynamics with good vertical thrust for drones of its class. The function of maintaining altitude and flips at 360o were performed without technical failures.

It is worth noting the load capacity of the Syma X8HG—with a cargo mass of 200 g, a significant loss of controllability and dynamics was not observed. With a cargo mass of 200–250 g, a gradual deterioration in the flight properties of the drone was observed.

Like other Syma models, the X8HG is not windproof due to its light weight, so it can only be used in calm weather. If used in conditions with a high wind speed, radio communication between the drone and the equipment is possible, followed by the unmanned vehicle leaving the radio and falling.

At the moment, the cost of Syma X8HG is about $120.

MJX BUGS 3. *The manufacturer of the quadcopter MJX BUGS 3* (Fig. 3) is the world-famous manufacturer of radio-controlled toys Meijiaxin Toys. The company is positioning this drone as such, designed both for aerial photography and for dynamic flights.

The features of this quadrocopter include:

- support for 3S batteries;
- control of battery charge and distance of flight;
- long flight time;
- axisless suspension;
- control at a frequency of 2.4 GHz;
- 360° flip;
- LED backlight.

Fig. 3 MJX BUGS 3

The quadcopter shell MJX BUGS 3 is made of nylon fiber, has established itself as a reliable and durable material, while the supports are made of ordinary plastic.

The quadcopter MJX BUGS 3 is equipped with MT1806 brushless motors with a power of 1800 kV. The manufacturer characterizes them as economical and efficient among the same type of brushless motors. Each motor provides traction of 230 grams.

Also in this quadrocopter are available ESC speed controllers with automatic anti-jamming, eliminating the possibility of motors burnout.

As mentioned above, complete with the drone there is an unknown suspension with the ability to manually adjust vertically, adapted for the installation of a small load. The distance from the ground to the suspension is 80 mm.

The quadcopter MJX BUGS 3 is powered by a 2S 1800 mAh Li-Po battery with a discharge current of 25C and an XT30 connector. According to the specification, the battery provides 19 min of continuous drone flight.

The quadrocopter also includes equipment operating at a frequency of 2.4 GHz. Its distinctive feature is the function of intelligent remote control, which reports a low battery charge or a large remoteness of the drone from the equipment. Its power is supplied using 4 AA batteries. The maximum distance of the drone from the control panel is 300–500 m.

During flight tests, the MJX BUGS 3 shows good flight performance even on the included 2S battery. The 6-axis gyroscope works without interruption. In practice, the flight time of the drone with a maximum load was 8 min. At a distance of 300 m, the quadrocopter clearly performs the specified flight directions.

The cost of the quadcopter MJX BUGS 3 in different stores is about $ 120.

Cheerson CX-20. The manufacturer of this model is the Chinese company Cheerson. The Cheerson lineup is presented in 3 versions: CX-10, SV-20 and SV-30

Fig. 4 Cheerson CX-20

According to company reports, the most popular type of quadrocopter is the Cheerson CX-20 Auto Pathfinder (Fig. 4).

The features of this quadrocopter include:

- intelligent orientation system;
- function "autopilot";
- open controller code;
- availability of GPS module;
- automatic return with landing in case of loss of connection or low battery;
- LED backlight.

The Chererson CX-20 quadrocopter can be equipped with 2 options of 2212 920 kV/1200 kV brushless motors.

Chererson CX-20 is equipped with a 3S Li-Po-battery with a capacity of 2700 mAh with a discharge current of 20–40C and a voltage of 11.1 V. During testing, this battery provided a continuous flight of the drone for 15 min without load for 10 min if available cameras. It takes 60 min to fully charge the battery.

The basic equipment of the Chererson CX-20 includes a 4-channel transmitter (equipment) operating at a 2.4 GHz frequency. It provides a range of up to 1 km in direct visibility. The transmitter is powered by 8 AA batteries.

Quadrocopter Cheerson CX-20 provides 3 versions of the flight controller:

1. "Open Source", thanks to the open code, allows operators to use the full potential of the ArduPilot controller and autopilot using the MissionPlanner software.

2. "ZERO" is developed on a 32-bit AMR STM32F103 RBT6 chip, there is no possibility to configure the controller, it is not compatible with the MissionPlanner software and accordingly has limited autonomous flight capabilities.
3. "Big Fly Shark" developed on the basis of the STM32F103 chip, the controller version is completed, it works with its own software.

Quadrocopter Cheerson CX-20 has 5 flight modes:

- Take Off—the mode of movement using the operator (without reference to GPS)
- GPS hold—position maintenance mode;
- Orientation—in this mode, the orientation of the drone in space is turned off;
- Altitude hold—height maintenance mode;
- Return-to-Home—the classic mode of return to the take-off point.

During flight tests, the Cheerson CX-20 showed good traction inherent to low-cost drones with brushless motors, and relative stability. Due to the large "windage", the quadrocopter is not recommended for use in windy weather.

The cost of the Chererson CX-20 quadrocopter in various online stores is from $ 250.

Hubsan X4 PRO H109S. The manufacturer of this quadrocopter is the Chinese company Shenzhen Hubsan Technology, which has long been producing unmanned aerial vehicles for domestic use. Compared to other models of the company, Hubsan X4 Pro H109S (Fig. 5) can be used for high-quality aerial photography.

The features of this quadrocopter include:

- different equipment;
- availability of GPS module

Fig. 5 Hubsan X4 Pro H109S

- automatic return during losing of communication;
- point flight mode;
- "Headless" mode;
- height maintenance function;
- FPV kit.

The HUBSAN X4 PRO quadrocopter is equipped with brushless motors, manufactured by Shenzhen Hubsan Technology.

The HUBSAN X4 PRO quadcopter is equipped with a high-performance lithium-polymer battery—3S with a capacity of 7000 mAh with a discharge current of 25C and a voltage of 11.1 V. According to the specification, this battery provides a drone flight time of 21 min. In practice, during using a mixed type of control, the battery lasts for 19 min of flight. Charging the battery lasts for 3 h.

The basic configuration of the HUBSAN X4 PRO includes an FPV-1 transmitter with an integrated LCD screen that displays video from the drone's camera, as well as flight telemetry and other important indicators, such as current coordinates, battery charge and remote control. This transmitter is immune to radio interference. The transmitter is powered by 8 AA batteries (can be replaced with a LiPo battery with a JST connector).

The HUBSAN X4 PRO quadcopter has 2 special flight modes:

- "Headless" mode—disables the orientation of the quadrocopter in space. It becomes unimportant for the pilot which side the drone is turned towards him;
- "Hold Altitude" mode—height maintenance mode. The integrated GPS module allows the drone to accurately maintain the position set by the pilot.

This quadrocopter also has 2 important functions:

- "Return-to-Home" function—the drone returns to the take-off point, followed by a soft landing, both automatically (in case of loss of communication), and from the control panel at the command of the pilot;
- "Flight to set points" function—the pilot sets the flight coordinates and altitude using the control panel, and after starting it participates only in the process of photo or video shooting.

The control range of the HUBSAN X4 PRO quadrocopter is up to 1 km, the flight height is 300–1000 m. The quadcopter's carrying capacity is about 200 g.

The cost of the HUBSAN X4 PRO quadrocopter ranges from $ 390 to $ 760 (depending on configuration).

Xiro Xplorer G. The manufacturer of this drone is the Chinese company Zero UAV Intelligance Technology Co. Ltd, is engaged in the design and development of highly intelligent unmanned systems for professional use, as well as small quadro-copters for domestic use. The company introduced a series of household quadro-copters XIRO XPLORER in 2015. One of the representatives of this series is the Xiro Xplorer G (Fig. 6).

The features of this quadrocopter include:

- shockproof modular design;

Fig. 6 Xiro Xplorer G

- variable clearance due to adjustable landing maggots;
- brushless power plant;
- GPS satellite positioning
- retention of position;
- autoslite and landing;
- flight mode by points (up to 16 points)
- "Return-to-Home" mode (RTH).

The quadrocopter Xiro Xplorer G is equipped with 2212 Zero UAV proprietary brushless motors. The developer does not specify information about the performance of motors. The power plant capacity is enough to realize a confident and stable flight even in windy weather.

The Xiro Xplorer G is powered by an intelligent 3S LiPo battery with an output voltage of 11.1 V and a capacity of 5200 mAh. The charging time is about 2 h. Such battery provides a maximum flight time of 25 min. The maximum horizontal flight speed is 28 km/h. The weight of the drone is reduced depending on the configuration within 1–1.3 kg.

Depending on the configuration, the Xiro Xplorer quadrocopter can be equipped with a 14 megapixel FHD camera, built on a 1/2.3-inch Panasonic CMOS matrix. The camera is placed on a 3-axis mechanical suspension.

The video stream is transmitted in real time using Wi-Fi at a frequency of 5.8 GHz. The video is displayed on the screen of the paired smartphone. According to the specification, the maximum removal distance is 800 m.

Xiro Explorer G is equipped with classic equipment operating at 2.4 GHz frequency. The equipment is powered by built-in lithium-ion battery.

Xiro Explorer G has two automatic flight modes:

- "Waypoints"—using a mobile application you can schedule automatic flight to points (up to 16 points maximum)

Fig. 7 3DR Solo

- "Return-to-Home" (RTH)—provides automatic drone return to take-off place when the battery is low;
- "Intelligent Orientation Control" (IOC)—an advanced analogue of the "Headless" mode, disabling the orientation of the drone in space and allows you to adapt to quadrocopter control.

The cost of the Xiro Explorer G quadrocopter is from $ 250 (depending on configuration).

3DR Solo. The manufacturer of this type of drone is the American company 3DRobotics. The main purpose of 3DR Solo (Fig. 7) is a professional highly intelligent and high-quality photo and video shooting.

This drone is controlled using a classic remote control connected to any gadget on iOS or Android through the 3DR SOLO APP application.

3DR Solo is equipped with brushless motors with a capacity of 880 kV, which is a proprietary of 3DRobotics.

The drone is equipped with a 5200 mAh LiPo battery with an output voltage of 14.8 V (DC). The battery is located at the top of the drone, and the battery charge time is about 1.5 h. The battery has a built-in LED indication of the state of charge.

3DR Solo proved to be excellent during flight tests: the maximum speed of the drone is 89 km/h, the speed of vertical ascent is up to 36 km/h, and the maximum ascent is 122 m (electronically limited). The flight time is about 20 min (with a payload of up to 15 min). The drone has excellent wind resistance. The maximum carrying capacity of the drone is 800 g. The weight of the drone is about 1.8 kg. The maximum distance of the drone from the control panel is 1 km.

The modular design of the 3DR Solo allows detailed replacement of damaged or defective ones. This will significantly reduce the cost of repairs after damage to the drone.

Quadrocopter 3DR Solo has 6 programmed modes:

- "Cable Cam"—allows you to autonomously move the drone at points specified before take-off;
- "Orbita"—allows the drone to move around a given point at an equal distance;
- "Selfy"—allows you to focus on the control panel and move to the specified height and distance;
- "Follow"—the mode of pursuit of the control panel;
- "Zipline"—a mode that provides autonomous flight exclusively along a direct path;
- "Pano"—a mode designed for photo and video shooting.

The cost of the 3DR Solo quadrocopter is from $ 400 (depending on configuration).
GoPro Karma. The manufacturer of this drone is a well-known company that manufactures GoPro cameras. The main objective of the company was the production of a drone (Fig. 8), it will be distinguished by its compactness, ease of use and the final quality of photos and videos.

GoPro Karma quadrocopter has 6 preset modes:

- "Follow"—allows the drone to automatically follow the user, keeping him in the center during the flight;
- "Watch"—the drone freezes and tracks the movement of the user with the equipment (the mode is implemented using GPS)
- "Dronie"—the mode is designed for wide-angle photography;
- "Cable Cam"—the mode is designed to move the drone at given points (up to 10 points)
- "Orbit"—allowing the drone to move around a given point at an equidistant distance;
- "Reveal"—a mode that provides autonomous flight along a direct path.

Fig. 8 GoPro Karma

The drone operates on the basis of a 4S LiPo battery with a capacity of 5100 mAh and an output voltage of 14 V. The time will help the battery is about 1 h. At the same time, the battery weighs 545 g. The battery is also equipped with an LED charge indication. Battery charging time is about 1 h.

The Karma GoPro traction is provided by 4 brushless motors, providing high flying characteristics of the drone. The maximum flight speed of GoPro Karma is 57 km/h, while the maximum flight altitude is 4.5 km.

The drone is able to withstand wind flows at speeds up to 10 km/h.

The maximum flight distance of the drone is 3 km, but in practice (in urban agglomerations) it does not exceed 700 m.

The GoPro Karma control panel is equipped with a high-resolution multifunction screen (720p). The charging time for the remote control is about 2.5 h.

The cost of a quadrocopter GoPro Karma is above $ 1000.

4 Results

This review includes 8 quadrocopters, which represent both the budget segment of the market (up to $ 300) and "premium" models, costing up to $ 1000. Naturally, this section criterion is conditional. During the review, the following quadrocopter parameters were highlighted, which are displayed in Table 1: model name, flight time (without payload), control range, altitude retention, camera availability, FPV,

Table 1 Technical specifications of quadrocopters

Quadcopter	Flight time, to min	Control range, to min	Altitude Hold	Camera included	FPV	Cost, from $
MJX X101	10	100	–	–	–	74
Syma X8HG	7–12	70	+	8Mp	–	120
MJX BUGS 3	8–19	500	–	–		119
Cheerson CX-20	15	1000	+	–	–	250
Hubsan X4 Pro H109S	25	1000	+	FHD	5.8 GHz	390
Xiro Explorer G	25	800	+	–	5.8 GHz (Wi-Fi)	237
3DR Solo	20	1000	+	–	3DR Link secure Wi-Fi network	399
GoPro Karma	20	3000	+	–	5.8 GHz	1060

28

A. Zaporozhets

Table 2 Advantages and disadvantages of quadrocopters

Quadcopter	Advantages	Disadvantages
MJX X101	design flight qualities vertical thrust Headless mode 3 activity modes suspension for cargo price	collector motors (small resource) lack of a camera in the kit
Syma X8HG	strength camera height maintenance function flight characteristics suspension for cargo Headless mode 6 axis gyroscope LED backlight	collector motors (small resource) additional balancing of the rotors is required
MJX BUGS 3	ease for using power feedback function at low charge and critical distance suspension for cargo compatible with 3S-batteries LED backlight price	lack of dynamism charged battery as standard
Cheerson CX-20	GPS module flight modes auto-return with landing (in case of loss of connection or low battery) flight range control LED backlight price	necessary to configure the controller no windproof lack of a camera in the kit
Hubsan X4 Pro H109S	GPS module flight range FPV kit/telemetry long flight time 3 complete sets	no "Follow me" function no auto take-off and auto-landing
Xiro Explorer G	modular design; flight characteristics; 3-axis suspension; battery with charge indication; automatic flight modes	sensitivity to radio interference fragile camera suspension fragile landing supports
3DR Solo	modular design; flight characteristics; hydro-stabilized 3-axis suspension; FPV OS Linux (drone/remote) unique flight modes	no able to fly around obstacles; adapted only for GoPro cameras lack of a camera in the kit

(continued)

Table 2 (continued)

Quadcopter	Advantages	Disadvantages
GoPro Karma	flight characteristics; stabilized 3-axis suspension; FPV flight modes camera location	lack of a camera in the kit no able to fly around obstacles flight time flight range cost

price. Table 2 contains the series of advantages and disadvantages of the considered quadrocopters.

5 Discussion

The analysis of the examined quadrocopters was carried out on the basis of the universal qualitative efficiency criterion (UQEC). This method is the simplest and most convenient for the task of choosing the optimal quadrocopter for monitoring in the field of energy and ecology. The UQEC indicates the achievement of the goal set for the monitoring tool.

In the considered problem, this criterion has the following boundary points: "10"—in the case of complete achievement of the goal; "0"—in case of complete failure of the goal. Thus, UQEC is the average value of particular qualitative efficiency criterion (PQEC) that characterize the achievement of individual goals set for a monitoring tool. In our task, they can also take a range of values from 0 to 10. The following parameters acted as PQEC: price/quality, camera, functionality, flight time, flight qualities, compactness, equipment, reliability. The values of PQEC are also the average value obtained as a result of a survey of users of the presented drones on the Dronomania website.

Figure 9 shows the values of PQEC for each individual parameter. Table 3 presents the results of particular performance criteria for each drone. "0" in the "Camera" column for many drones implies the absence of a camera in the basic set of the drone.

In Fig. 10, as well as in Table 3, the UQEC values for each quadrocopter are displayed. According to the results obtained, quadrocopters can be divided into 2 groups: with a relatively low UQEC (ranging from 6 to 7) and relatively high UQEC (in the ranges from 8 to 9). None of the examined quadrocopters has the maximum value of the UQEC. Such models as MJX X101, Syma X8HG, MJX BUGS 3, Cheerson CX-20 were included in the group with relatively low UQEC. These quadrocopters are characterized by a relatively cheap cost—from 75 to 250 $.

Such models as Hubson X4 Pro H109S, Xiro Explorer G, 3DR Solo, GoPro Karma fell into the group with relatively high UQEC. These quadrocopters are characterized by a relatively high cost—from $ 250 to $ 1,000.

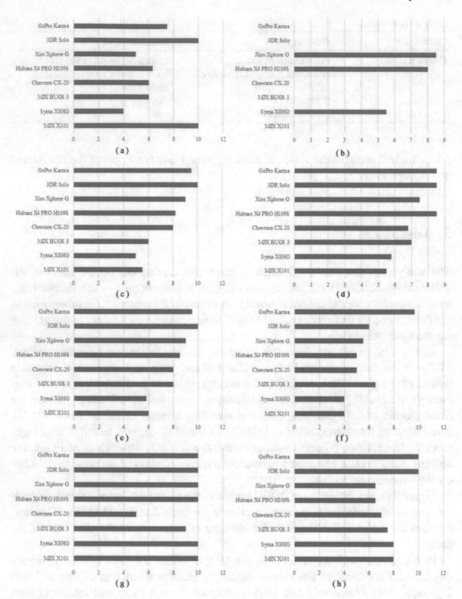

Fig. 9 Values of particular qualitative efficiency criterion for quadrocopters for each parameter: **a** price/quality; **b** camera; **c** functionality; **d** flight time; **e** flight qualities; **f** compactness; **g** equipment; **h** reliability

Table 3 PQEC and UQEC of quadrocopters

Quadcopter	Price/quality	Camera	Functionality	Flight time	Flight qualities	Compactness	Equipment	Reliability	UQCE
MJX X101	10	0	5.5	5.5	6	4	10	8	6.13
Syma X8HG	4	5.5	5	5.8	5.8	4	10	8	6.01
MJX BUGS 3	6	0	6	7	8	6.5	9	7.5	6.25
Cheerson CX-20	6	0	8	6.8	8	5	5	7	5.73
Hubsan X4 PRO H109S	6,3	8	8.2	8.5	8.5	5	10	6.5	7.63
Xiro Xplorer G	5	8.5	9	7.5	9	5.5	10	6.5	7.63
3DR Solo	10	0	10	8.5	10	6	10	10	8.06
GoPro Karma	7.5	0	9.5	8.5	9.5	9.7	10	10	8.09

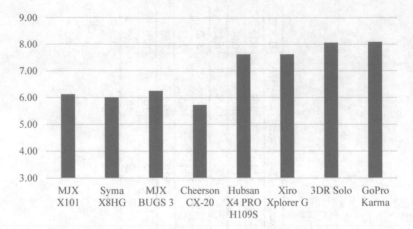

Fig. 10 UQEC of quadrocopters

The quadcopter with the lowest UQEC is MJX X101 (6.13), the quadrocopter with the highest UQEC is GoPro Karma (8.40).

Of course, for various monitoring tasks, one of the considered parameters can be critically important. In this case, a quadrocopter with such parameters should be used. However, based on the analysis, it can be concluded that the MJX BUGS 3 quadrocopter has the highest UQEC rating among the quadrocopters under consideration with respect to the relatively inexpensive market segment. Its application for monitoring tasks in the field of energy and ecology is justified.

6   Practical Implementation

To perform experiments with thermal imaging of heat supply pipelines on the basis of the MJX BUGS 3, a compact thermal imaging camera manufactured by Seek Thermal ™ (USA) was installed, which has a wide-angle lens with a total size of 2.5 × 4.4 × 2.5 cm resolution of 320 × 240. The greatest shooting distance is 610 m, and the closest distance is 15 cm. The appearance of the diagnostic system of main pipelines of heat networks based on MJX BUGS 3 is shown in Fig. 11.

For monitoring extended objects (in our case, pipelines of heat networks) it is proposed to fly around the test object using a multi-rotor type UAV. This allows you to get high-quality photos and thermal images of the site of the heating system as an object of control for further analysis. The software allows you to use any topographic basis as a map. Binding can be done at two or more points. It is also possible to use as a topological basis of electronic maps. The program provides input, automatic control and editing of the route of the flight. An elevation can be specified for each waypoint.

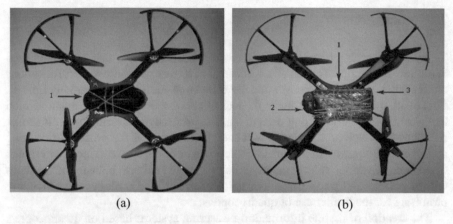

(a)   (b)

Fig. 11 Hardware complex for monitoring the technical condition of main pipelines: **a** top view; **b** bottom view; 1—UAV, 2—thermal imaging camera; 3—communication device

The results of measuring the thermal state of the heat grids were carried out using a thermal imaging camera, mounted on a UAV [22].

Figure 12 shows thermal images of sections of the heat network where experimental studies were conducted. The shooting was performed on November 22, 2018 at 18 p.m. on a cloudless sky at an air temperature of –4 °C.

Figure 12 clearly shows the possibility of identifying defective areas of main pipelines using low power quadrocopters.

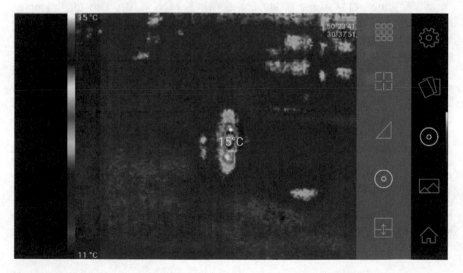

Fig. 12 The working environment of the software complex based on Seek Thermal

7 Conclusions

Today we can state the rapid development of quadrocopters, which are mainly used in military operations. The list of subject areas of the use of quadrocopters for various other studies, operations, conventionally called non-military, is essentially limited. First of all, this limitation is due to the lack of created, developed and manufactured technical tools for conducting diverse studies, primarily measuring tools. It can be predicted that such an imbalance will soon be broken and the process of creating appropriate equipment for the quadrocopters will be adjusted to conduct a wide range of research in various subject areas, among which the defense industry of the states will be priority and prospective. The fields of energy and ecology are particularly promising for the further use of quadrocopters.

The creation of mobile information-measuring systems based on quadrocopters makes it possible to diagnose the state and dynamics of the characteristics in time and space of the studied environment, both in on-line modes and other modes. The on-line mode is especially effective in case of accidents in areas of spatially branched heat networks. With normal operation of the objects under study, the current remote control is the most economical compared with other means of control. This allows to use such measuring tools to create the necessary databases for diagnostics the characteristics of the thermal state of heat networks to predict their dynamics.

This review of serial quadrocopters provides a wide array of data on their technical capabilities. The analysis showed promising quadrocopter models that can be used for monitoring in the fields of energy and ecology. Their distinguishing feature among other production models is the presence of a suspension and the ability to lift loads from 250 grams. One of the promising models is MJX X101. This quadrocopter, according to its users, is well equipped, reliable and has a relatively low cost. The best model in the result of the analysis was GoPro Karma. This model showed one of the best results for the selected evaluation criteria.

In general, the use of quadrocopters to solve monitoring tasks in the fields of ecology and energy will be developed in the future. Compact quadrocopters have high flight characteristics and low cost, which makes them competitive in the field of monitoring.

Acknowledgements The project presented in this article is supported by ≪Development of a system for monitoring the level of harmful emissions of TPP and diagnosing the equipment of power plants using renewable energy sources on the basis of Smart Grid with their collaboration≫ (2019–2021, 0119U101859), which are financed by National Science of Ukraine, and ≪Development of a system for monitoring micro climatic parameters and the air pollution of the ecosystems the Northern Black Sea Coast≫ (2019–2021, 0119U100550), which is financed by Ukrainian Ministry of Education.

References

1. Goudie, A.S.: Human Impact on the Natural Environment. John Wiley & Sons (2018)
2. Fuss, S., Jones, C.D., Kraxner, F., Peters, G.P., Smith, P., Tavoni, M., van Vuuren, D.P., Canadell, J.G., Jackson, R.B., Milne, J.: Research priorities for negative emissions. Environ. Res. Lett. **11**(11), 115007 (2016). https://doi.org/10.1088/1748-9326/11/11/115007
3. Babak, V.P., Babak, S.V., Myslovych, M.V., Zaporozhets, A.O., Zvaritch, V.M.: Simulation and software for diagnostic systems. In: Diagnostic Systems For Energy Equipments. Studies in Systems, Decision and Control, vol. 281, pp. 71–90. Springer, Cham (2020). https://doi.org/10.1007/978-3-030-44443-3_3
4. Agarwal, A., Shukla, V., Singh, R., Gehlot, A., Garg, V.: Design and development of air and water pollution quality monitoring using IoT and quadcopter. In: Singh R., Choudhury S., Gehlot A. (eds.) Intelligent Communication, Control and Devices. Advances in Intelligent Systems and Computing, vol. 624, pp. 485–492. Springer, Singapore (2018). https://doi.org/10.1007/978-981-10-5903-2_49
5. Zaporozhets, A., Eremenko, V., Isaenko, V., Babikova, K.: Approach for creating reference signals for detecting defects in diagnosing of composite materials. In: Shakhovska N., Medykovskyy M. (eds.) Advances in Intelligent Systems and Computing IV. CCSIT 2019. Advances in Intelligent Systems and Computing, vol. 1080, pp. 154–172, Springer, Cham (2020). https://doi.org/10.1007/978-3-030-33695-0_12
6. Gageik, N., Reul, C., Montenegro, S.: Autonomous quadrocopter for search, count and localization of objects. In: Guanghui W. (ed.) Recent Advances in Robotic Systems, InTech, pp. 1–25 (2016). https://doi.org/10.5772/63568
7. Skorupka, D., Duchaczek, A., Waniewska, A.: Optimisation of the choice of UAV intended to control the implementation of construction projects and works using the AHP method. Czasopismo Techniczne **9**, 117–125 (2017). https://doi.org/10.4467/2353737XCT.17.152.7164
8. Oats, R.C., Escobar-Wolf, R., Oommen, T.: Evaluation of photogrammetry and inclusion of control points: significance for infrastructure monitoring. Data **4**(1), 42 (2019). https://doi.org/10.3390/data4010042
9. Anweiler, S., Piwowarski, D.: Multicopter platform prototype for environmental monitoring. J. Cleaner Prod. **155**(1), 204–211 (2017). https://doi.org/10.1016/j.jclepro.2016.10.132
10. Kharchenko, V., Sachenko, A., Kochan, V., Fesenko, H.: Reliability and survivability models of integrated drone-based systems for post emergency monitoring of NPPs. In: 2016 International Conference on Information and Digital Technologies (IDT), July 5–7, pp. 127–132. Rzeszow, Poland (2016). https://doi.org/10.1109/dt.2016.7557161
11. Babak, S., Babak, V., Zaporozhets, A., Sverdlova, A.: Method of statistical spline functions for solving problems of data approximation and prediction of objects state. In: CEUR Workshop Proceedings, vol. 2353, pp. 810–821 (2019). http://ceur-ws.org/Vol-2353/paper64.pdf
12. Roger, M., Prahl, C., Pernpeinter, J., Sutter, F.: 7—new methods and instruments for performance and durability assessment. In: The Performance of Concentrated Solar Power (CSP) Systems, pp. 205–252. Woodhead Publishing (2017). https://doi.org/10.1016/b978-0-08-100447-0.00007-9
13. Nishar, A., Richards, S., Breen, D., Robertson, J., Breen, B.: Thermal infrared imaging of geothermal environments and by an unmanned aerial vehicle (UAV): a case study of the Wairakei—Tauhara geothermal field, Taupo, New Zealand. Renewable Energy, **86**, 1256–1264 (2016). https://doi.org/10.1016/j.renene.2015.09.042
14. Nishar, A., Richards, S., Breen, D., Robertson, J., Breen, B.: Thermal infrared imaging of geothermal environments by UAV (unmanned aerial vehicle). J. Unmanned Veh. Sys. **4**(2), 136–145 (2016). https://doi.org/10.1139/juvs-2015-0030
15. Tian, Y., Lo, D., Xia, X., Sun, C.: Automated prediction of bug report priority using multi-factor analysis. Empirical Softw. Eng. **20**(5), 1354–1383 (2015). https://doi.org/10.1007/s10664-014-9331-y

16. Tian, Y., Lo, D., Sun, C.: DRONE: predicting priority of reported bugs by multi-factor analysis. In: 2013 IEEE International Conference on Software Maintenance, Sept. 22–28, pp. 200–209. Eindhoven, Netherlands (2013). https://doi.org/10.1109/icsm.2013.31
17. Babak, V.P., Babak, S.V., Myslovych, M.V., Zaporozhets, A.O., Zvaritch, V.M.: Technical provision of diagnostic systems. In: Diagnostic Systems For Energy Equipments. Studies in Systems, Decision and Control, vol. 281, pp. 91–133. Springer, Cham (2020). https://doi.org/10.1007/978-3-030-44443-3_4
18. Babak, V.P., Babak, S.V., Myslovych, M.V., Zaporozhets, A.O., Zvaritch, V.M.: Principles of construction of systems for diagnosing the energy equipment. In: Diagnostic Systems For Energy Equipments. Studies in Systems, Decision and Control, vol. 281, pp. 1–22. Springer, Cham (2020). https://doi.org/10.1007/978-3-030-44443-3_1
19. Jung, N.-M., Choi, M.-H., Lim, C.-W.: Development of drone operation system for diagnosis of transmission facilities. In: 2018 21st International Conference on Electrical Machines and Systems (ICEMS), Oct. 7–10, pp. 2817–2821. Jeju, South Korea (2018). https://doi.org/10.23919/icems.2018.8549507
20. Yang, X., Du, J., Liu, S., Li, R., Liu, H.: Air pollution source estimation profiling via mobile sensor networks. In: 2016 International Conference on Computer, Information and Telecommunication Systems (CITS), July 6–8. Kunming, China (2016). https://doi.org/10.1109/cits.2016.7546456
21. Lee, K.-B., Kim, Y.-J., Hong, Y.-D.: Real-time swarm search method for real-world quadcopter drones. Appl. Sci. **8**(7), 1169 (2018). https://doi.org/10.3390/app8071169
22. Zaporozhets, A., Kovtun, S., Dekusha, O.: System for monitoring the technical state of heating networks based on UAVs. In: Shakhovska, N., Medykovskyy, M. (eds.) Advances in Intelligent Systems and Computing IV. CCSIT 2019. Advances in Intelligent Systems and Computing, vol. 1080, pp. 935–950. Springer, Cham (2020). https://doi.org/10.1007/978-3-030-33695-0_61

Researches of the Stressed-Deformed State of the Power Structures of the Plane

Fomichev Petr⑩, Zarutskiy Anatoliy⑩, and Lyovin Anatoliy⑩

Abstract The paper describes a method for determining the thrust of an aircraft engine according to tensometry of the pylon power structure, which is implemented on a non-flying aircraft. A key feature of this method is the combination of strain gauge method and the calculated definition of deformations in the power elements of the structure using the finite element method. A test bench was designed and installed in the static test hall, which includes the aircraft pylon, the systems for fastening and loading the structure, and the tensometric system. The stress-strain state of the pylon structure has been analyzed in order to select rational places for sticking strain gauges. A geometric model of the pylon structure was constructed to create its finite element model. Deformations were determined at tensometric points under different conditions of application of maximum thrust and reverse using a calculation model in a finite element package. Pylon was tested with a load of up to 12,500 kg, simulating the effect of maximum thrust. The assessment of the adequacy of the computational model was carried out by comparing the results of the calculation of deformations and strain gauge data at the control points of the structure.

Keywords Engine thrust · Gas turbine engines · Plane · Pylon · Strain gauge sensors · Simulation

F. Petr · Z. Anatoliy (✉)
M.E. Zhukovsky National Aerospace University, Kharkiv Aviation Institute, Kharkiv, Ukraine
e-mail: zarutskiyanatoliy@gmail.com

L. Anatoliy
Concern "Titan", Kyiv, Ukraine

V. Babak et al. (eds.), *Systems, Decision and Control in Energy I*, Studies in Systems,
Decision and Control 298, https://doi.org/10.1007/978-3-030-48583-2_3

1 Introduction

A significant resource of modern aircraft engines requires constant monitoring of their condition during operation. Along with the traditional methods of periodically monitoring of the engine's state (by standard parameters, visual, optical, etc.), they use integrated methods based on the control of engine thrust and determine the tendency of its change during operation.

The methods for determining the thrust of gas turbine engines (GTE) at flying laboratories and as part of an operating aircraft are divided into: (1) gas-dynamic; (2) aerodynamic; (3) dynamometric; (4) using probes at the nozzle exit.

Also, for measuring the thrust of aircraft engines, gas-dynamic stands are used, which are placed in special rooms (laboratories). Such an approach requires the introduction of corrections for the measured thrust associated with air intake, an inlet device, and gas exhaust. Correction values are set based on the equations of gas dynamics and additional calibrations.

The methods presented above are implemented to a greater extent on ground test benches, and cannot be used on an ongoing basis in aircraft operation. The value of engine thrust is the most important parameter during the flight, which displays the qualitative characteristics of the aircraft and the engine as a whole.

Monitoring the state of the aircraft engine that is in operation is very important. There are quite a few power equipment monitoring systems that use multi-sensor systems. In [1, 2], methods and a system for diagnosing heat power equipment that have multifunctional use are displayed. Also, indirect methods for calculating parameters are used for diagnosis, which are applicable for monitoring the parameters of on-board systems [3, 4]. In some cases, methods of virtual maintenance and basic ergonomics are applicable [5].

To control the parameters of aircraft structural elements, an analysis of the stress-strain state is used [6, 7]. Detection of unstable operating modes of structural elements is carried out using vibrations [8, 9]. A method for changing the concentration of oxygen in air was presented in [10, 11], which can also be applied for measuring engine thrust to eliminate a methodological error.

The developed information-analytical systems that are implemented as laboratories on airplanes [12, 13] are of considerable interest. In some cases, the test bench may be suspended on springs to an aircraft engine [14]. In [15], a system is described that implements the measurement of the stress-strain state of pylon structures based on strain gauge sensors, but the authors do not use it to measure the thrust of an aircraft engine.

Thus, there is an urgent task of determining the thrust of an aircraft engine during operation.

2 Method and Experimental Equipment for Measuring Engine Thrust

2.1 Aircraft Engine Thrust Measurement Method

Measurement of engine thrust carried out on an experimental stand. In order to eliminate the need to calibrate the thrust measurement system after installing the engine on an airplane, it is proposed to perform graduation in the static test hall.

The measurement system includes strain gauge sensors glued at the strain gauge points on the main power elements of the pylon, a strain gauge station installed permanently in the engine mount pylon, and switching wires. The strain gauge station includes a power supply, analog signal amplifiers, and an analog-to-digital converter. A digital signal proportional to the strain at the strain gauge point is supplied to the computing device.

The position of the strain gauge points is selected based on the preliminary calculation of the stress-strain state of the pylon by the finite element method.

The essence of the proposed determination method is in follows.

At the first stage, the calibration of the pylon in the hall of static tests is carried out. The dependence of the strains ε_p at the strain gauge points on the load P applied to the vertical pin of the front node of the engine mount on the pylon is established. Through this pin, the engine thrust is transmitted to the aircraft structure.

Next, the strains ε_p are calculated by the finite element method at the tensometric points of the pylon structure during a load P is applied to the pin of the front engine mount assembly. The calculation of strains ε_T is carried out by the finite element method at the tensometric points of the pylon structure during a load T is applied along the axis of symmetry of the geometric layout of the engine. The thrust conversion factor η is defined as the ratio of deformations at the tensometric points of the pylon structure during loads are applied along the axis of the engine T to the pin of the front linkage assembly P $(P = T)$. Thrust measurement is carried out with the engine running on the experimental stand in the aircraft system. Further, structural deformations at tensometric points are determined, taking into account the calibration curve, and using the thrust conversion factor η, it find the engine thrust.

2.2 Pylon Loading Stand and Strain Gauge Equipment

To carry out the calibration of strain gauge points in the power structure of the pylon, a test bench was developed and mounted in the static testing hall (Strength Laboratory, Kharkov Aviation Institute). The stand includes the IL-76 airplane pylon, fastening and structural loading systems, and a strain gauge system. The test bench layout is shown in Fig. 1.

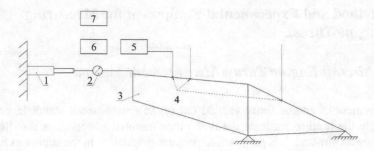

Fig. 1 Scheme of the test bench: **1**—hydraulic cylinder; **2**—dynamometer; **3**—plane's pylon; **4**—strain gauge points; **5**—measuring transducer; **6**—power supply with interface converter; **7**—computer

Given the design features of the static testing hall, the pylon in the stand is installed upside down. The fixing on the pylon attachment points to the wing is made to the power floor using transition nodes.

The loading system for simulating engine thrust includes a hydraulic cylinder with a manual hydraulic pump station and a dynamometer (DPU-20-2-U2, GOST 13837-68) to control the magnitude of the load. Before the test, a metrological verification of the dynamometer was carried out by a specialist of the Kharkov Institute of Metrology with the issuance of the appropriate approval for the test.

A general view of the test bench is shown in Fig. 2.

Strain gages (KF 5P1-3-100-B-12) were used to measure strains at strain points, the principle of operation of which is based on the property of the grating conductor to change the ohmic resistance due to tension or compression.

For gluing the strain gages on metals and non-porous materials, one-component cold hardening adhesive (cyacrine) was used, recommended by the manufacturer of strain gages.

3   Simulation of the Design of an Airplane Pylon

An analysis of the stress-strain state of the structure was carried out to determine the most rational places for sticking the strain gauges to the power elements of the pylon. The calculations were performed by the finite element method in a geometrically and physically linear formulation.

To create a finite element model, a geometric model of the pylon structure was preliminarily constructed (Fig. 3).

The geometric parameters (overall dimensions, shape and dimensions of cross sections, thicknesses) were taken from drawing 1.7601.2010.200.000 "General view of the caisson of the internal engine pylon" and specified on the full-scale structure of the pylon. The power part of the pylon is considered: the upper and lower beams, between which 17 frame ribs, a strut beam, skin, fairing are installed.

Fig. 2 General view of the test bench for measuring aircraft engine thrust

The preference is given to a planar model, since planar finite elements are well suited for modeling the structure under study, and their number in the model will be much less than the number of volume elements, necessary for modeling such structure.

On the basis of a planar geometric model, a finite element (FE) grid is constructed (Fig. 4). The grid is built in semi-automatic mode. For lining and ribs, flat octagonal finite elements (Shell type) of a quadrangular shape and, in places with complex geometry, six-node triangular shapes are used. For belts and lower beams, as well as strut beams, linear finite elements of the Beam type were used.

The FE grid obtained by the semi-automatic method does not have any pronounced violations of the technology, incorrect elements, or other possible shortcomings of the automatically generated grid.

Material properties are set using elastic characteristics—elastic modulus and Poisson's ratio.

The conditions for fixing and loading the pylon model are as close as possible to those that are implemented in a real design.

The system of fastening the engine to the pylon consists of 3 nodes: (1) front central (power) fastening node—takes the load from the engine rod transmitted through a spherical bearing to the vertical pin of the pylon; (2) front lateral fasteners—perceive part of the load from the weight of the engine and torque transmitted through traction; (3) rear attachment site—perceives part of the load from the weight of the engine

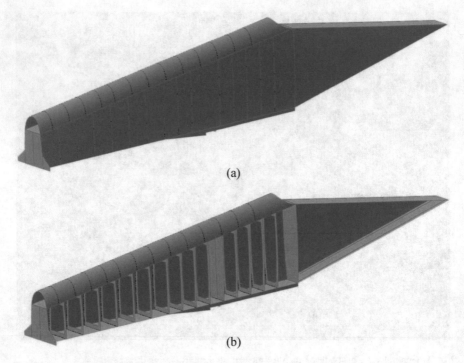

(a)

(b)

Fig. 3 Geometric model of the pylon structure: **a**—general view; **b**—inside view

and lateral loads transmitted through massive traction. The articulation of the thrust compensates for the expansion of the engine structure due to heating.

The finite element model is pivotally fixed at 4 points simulating the attachment points of the pylon to the wing of the aircraft. The engine is modeled as a rigid body with a given weight. The simulation of the attachment points to the pylon is made using the imposition of relations:

- a ban on vertical movements is imposed on the rear attachment point and front side attachment points;
- horizontal front mounting (pin) prohibited horizontal movement (in flight);
- a ban on lateral movements is also imposed on the rear attachment point.

4   Results of Simulation of Pylon's Loading

To assess the adequacy of the design model of the pylon, the calculation of the stress-strain state is performed for the case when the force is applied to the pin.

Two cases are considered: (1) direct thrust of 8000 kg; (2) reverse thrust 4000 kg.

As a result, the distribution of deformations along the belts of the lower beam is obtained.

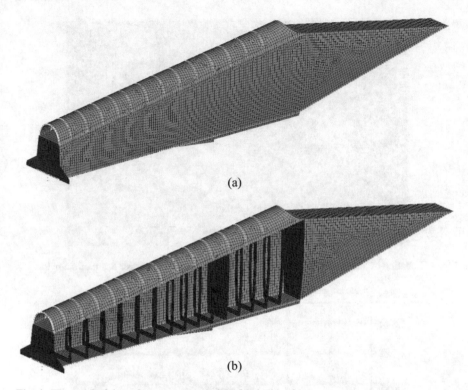

(a)

(b)

Fig. 4 FE model of the pylon: **a**—general view; **b**—inside view

Comparison of the calculation results is performed with experimental data. For this purpose, temporary tensometric sensors are glued to the belts of the lower beam at three tensometric points and strain measurements are carried out.

Comparison of the calculation results is performed with experimental data. For this purpose, temporary tensometric sensors are glued to the belts of the lower beam at three tensometric points and strain measurements were performed (Fig. 5).

Figure 6 shows a comparison of the calculation results with the strain gauge data of the pylon structure. The origin corresponds to the position of the vertical pin of the front engine mount on the pylon. The jumps on the diagrams are associated with a local change in the areas of the longitudinal belts of the lower beam.

As a result, it was found that the strain gauge data of the structure are in satisfactory agreement with the calculation results.

In order to determine the places of sticker for strain gauge sensors, a study was made of the stress-strain state of the pylon structure under the action of maximum traction (direct and reverse) and taking into account the weight of the engine. The load from the thrust is applied along the axis of the engine at the center of gravity.

Figure 7 shows the distribution of the total displacements in the pylon structure under the action of an engine thrust of 12,000 kg.

Fig. 5 Location of temporary strain gauge sensors on the belt of the lower beam of the pylon

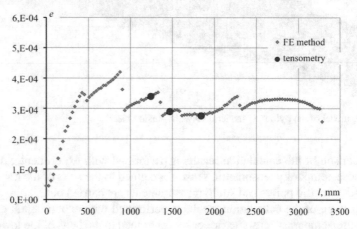

Fig. 6 Comparison of calculation results with pylon strain gauge data for direct engine thrust

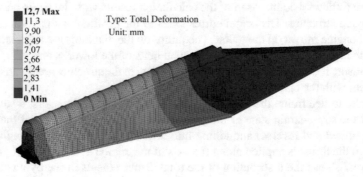

Fig. 7 Distribution of total displacements

The analysis of the structural-power scheme of the pylon showed that as the main power elements for the sticker of strain gauge sensors, the belts of the upper and lower beams, as well as the strut beam, can be considered. The criteria for choosing the places of the sticker of the sensors adopted the value of longitudinal deformation and ease of installation.

Figures 8, 9 and 10 show the distribution of longitudinal deformation along the lengths of the belts of the lower and upper beams, as well as along the strut beam (direct thrust). On the graphs, the zone is highlighted in a different color, where there is an opportunity for the sensors to be sticker (hatches, panels removed).

A similar analysis was carried out for the case of reverse engine thrust (4000 kg).

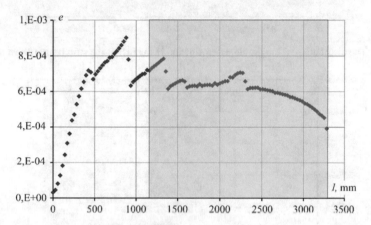

Fig. 8 The distribution of longitudinal deformation in the belt of the lower beam (direct thrust)

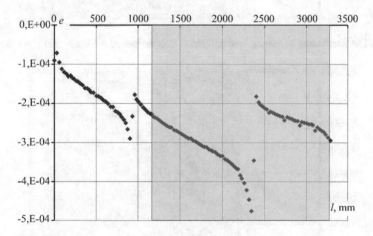

Fig. 9 The distribution of longitudinal deformation in the belt of the upper beam (direct thrust)

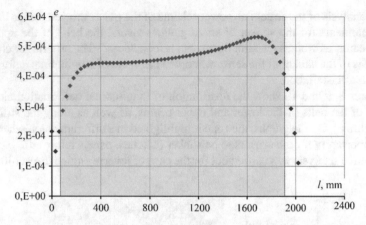

Fig. 10 The distribution of longitudinal deformation in the belt of the strut beam (direct thrust)

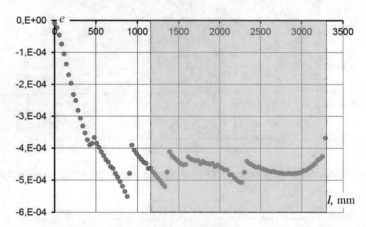

Fig. 11 The distribution of longitudinal deformation in the belt of the lower beam (reverse thrust)

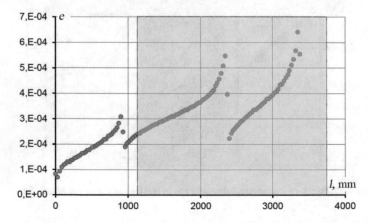

Fig. 12 The distribution of longitudinal deformation in the belt of the upper beam (reverse thrust)

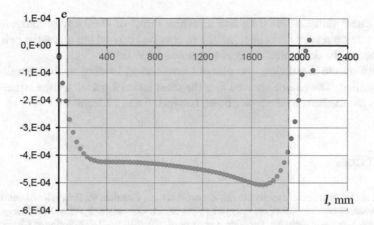

Fig. 13 The distribution of longitudinal deformation in the belt of the strut beam (reverse thrust)

Table 1 Coordinates of the position of strain gauge points	Strain point number	The coordinate along the length of the belt from the pin, mm
	№ 1	1245
	№ 2	1840
	№ 3	1470
	№ 4	1470

Figures 11, 12 and 13 show the dependences of the distribution of longitudinal deformation along the length of the belts of the lower and upper beams, as well as along the strut beam (reverse thrust).

Based on the results of the calculations, it was found that the most rational places for the strain gauge stickers are the lower beam belts between 6 and 17 frames (there is an approach due to removable panels). Table 1 shows the coordinates of the position of the strain gauge points along the length of the belt. The reference point is the vertical pin of the front engine mount on the pylon.

5 Conclusions

Using the finite element method, a 3D model of the pylon of the IL-76 aircraft was created. It is based on flat eight-node finite elements of the Shell type of a quadrangular shape, and in places with a significant number of geometric changes— six-node finite elements of a triangular shape.

The analysis of the stress-strain state of the pylon structure based on the created 3D model is performed. The adequacy of the design model of the pylon was tested on the basis of an assessment of the stress-strain state of the pylon when it is loaded with a direct thrust of 8000 kg and a reverse thrust of 4000 kg. A comparison of the simulation results and direct measurements was carried out along the belt of the lower beam, the determination coefficient was $R^2 = 0.99$.

Based on the simulation, the places of maximum loading of the aircraft pylon are identified. The results showed that the most rational places for the strain gauge stickers are the belts of the lower beam between 6 and 17 frames.

References

1. Babak, V., Babak, S., Myslovuch, M., Zaporozhets, A., Zvaritch, V.: Principles of construction of systems for diagnosing the energy equipment. In: Diagnostic Systems For Energy Equipments. Studies in Systems, Decision and Control, vol. 281, pp. 1–22. Springer, Cham (2020). https://doi.org/10.1007/978-3-030-44443-3_1
2. Zaporozhets, A., Eremenko, V., Isaenko, V., Babikova, K.: Approach for creating reference signals for detecting defects in diagnosing of composite materials. In: Shakhovska, N., Medykovskyy, M. (eds.) Advances in Intelligent Systems and Computing IV. CCSIT 2019. Advances in Intelligent Systems and Computing, vol. 1080, pp. 154–172. Springer, Cham (2020). https://doi.org/10.1007/978-3-030-33695-0_12
3. Kanyshev, A.V., Korsun, O.N., Ovcharenko, V.N., Stulovskii, A.V.: Identification of aerodynamic coefficients of longitudinal movement and error estimates for onboard measurements of supercritical angles of attack. J. Comput. Sys. Sci. Int. **57**(3), 374–389 (2018). https://doi.org/10.1134/S1064230718030048
4. Babak, S., Babak, V., Zaporozhets, A., Sverdlova, A.: Method of statistical spline functions for solving problems of data approximation and prediction of objects state. In: CEUR Workshop Proceedings, vol. 2353, pp. 810–821 (2019). http://ceur-ws.org/Vol-2353/paper64.pdf
5. Xu, H.C., Pan, Y., Xiong, Y.X., He, J.W.: Comparative study on two civil engine thrust reversers for their maintainability analysis. In MATEC Web of Conferences, vol. 169, p. 01026. EDP Sciences (2018). https://doi.org/10.1051/matecconf/201816901026
6. Fomichev, P.A.: Method for the evaluation of the service life under random loading based on the energy criterion of fatigue fracture. Strength Mater. **40**(2), 224–235 (2008). https://doi.org/10.1007/s11223-008-9011-5
7. Fomichev, P.A.: Substantiation of the predicted fatigue curve for structural elements of aluminum alloy. Strength Mater. **43**(4), 363 (2011). https://doi.org/10.1007/s11223-011-9305-x
8. Babak, V., Filonenko, S., Kornienko-Miftakhova, I., Ponomarenko, A.: Optimization of signal features under object's dynamic test. Aviation **12**(1), 10–17 (2008). https://doi.org/10.3846/1648-7788.2008.12.10-17
9. Chen, G., Rui, X., Yang, F., Zhang, J.: Study on the natural vibration characteristics of flexible missile with thrust by using Riccati transfer matrix method. J. Appl. Mech. **83**(3), 031006 (2016). https://doi.org/10.1115/1.4032049
10. Zaporozhets, A.: Research of the process of fuel combustion in boilers. In: Control of Fuel Combustion in Boilers. Studies in Systems, Decision and Control, vol. 287, pp. 35–60. Springer, Cham (2020). https://doi.org/10.1007/978-3-030-46299-4_2
11. Babak, V.P., Babak, S.V., Myslovych, M.V., Zaporozhets, A.O., Zvaritch, V.M.: Technical provision of diagnostic systems. In: Diagnostic Systems For Energy Equipments. Studies in Systems, Decision and Control, vol. 281, pp. 91–133. Springer, Cham (2020). https://doi.org/10.1007/978-3-030-44443-3_4
12. Mirabbashi, A.S., Mazidi, A., Jalili, M.M.: Analytical and experimental flutter analysis of a typical wing section carrying a flexibly mounted unbalanced engine. Int. J. Struct. Stab. Dyn. **19**(02), 1950013 (2019). https://doi.org/10.1142/S0219455419500135

13. Bauer, M., Friedrichs, J., Wulff, D., Werner-Spatz, C.: Development and validation of an on-wing engine thrust measurement system. In ASME Turbo Expo 2017: Turbomachinery Technical Conference and Exposition. American Society of Mechanical Engineers Digital Collection (2017, June). https://doi.org/10.1115/GT2017-63277

14. Zhu, H., Du, F., Zhao, S., Xu, Z., Lv, Z.: Design of a new type thrust measuring system for micro-turbojet engine. In 2016 2nd International Conference on Artificial Intelligence and Industrial Engineering (AIIE 2016). Atlantis Press (2016, November). https://doi.org/10.2991/aiie-16.2016.109

15. Bauer, M., Friedrichs, J., Wulff, D., Werner-Spatz, C.: Measurement quality assessment of an on-wing engine thrust measurement system. In ASME Turbo Expo 2018: Turbomachinery Technical Conference and Exposition. American Society of Mechanical Engineers Digital Collection (2018, June). https://doi.org/10.1115/GT2018-76496

Improving Method for Measuring Engine Thrust with Tensometry Data

Fomichev Petr⬡, Zarutskiy Anatoliy⬡, and Lyovin Anatoliy⬡

Abstract The paper describes a method for determining the thrust of an aircraft engine according to tensometry of the pylon's power structure, which is implemented on a flightless aircraft. A key feature of this method is the combination of the method of tensometry and the calculated determination of deformations in the structural elements using the finite element method. This approach allows the calibration of the measurement system in stationary conditions of the hall of static tests, eliminating the need to create a prototype engine for applying traction along its axis. Due to the fact that the indications of some strain gauge sensors during the engine was running on a stand based on the IL-76 aircraft had differences from the laboratory ones, the influence of local temperature changes from the working engine on the stability of the sensor readings was analyzed. New places for sticking strain gauge sensors have been selected, eliminating the effect of elevated temperature during the engine is running. According to the results of the calibration of the pylon in the hall of statistical tests at each tensometric point, the dependences of the deformations on the applied load, simulating the engine thrust from 0 to 12 tons, are obtained. New calibration coefficients are established, which are included in the software for processing thrust measurements.

Keywords Engine thrust · Aircraft · Tensometry · Strain gauge sensors · Reactive moment

F. Petr · Z. Anatoliy (✉)
M.E. Zhukovsky National Aerospace University, Kharkiv Aviation Institute, Kharkiv, Ukraine
e-mail: zarutskiyanatoliy@gmail.com

L. Anatoliy
Concern "Titan", Kyiv, Ukraine

V. Babak et al. (eds.), *Systems, Decision and Control in Energy I*, Studies in Systems, Decision and Control 298, https://doi.org/10.1007/978-3-030-48583-2_4

51

1   Introduction

The methods for determining the thrust of gas turbine engines (GTE) at flying laboratories and as part of an operating aircraft are divided into: (1) gas-dynamic; (2) aerodynamic; (3) dynamometric.

The gas-dynamic method is the main method for determining the thrust of a GTE in high-speed conditions. The method is used to study the flight characteristics of all domestic engines in the flying laboratories Tu-16LL and IL-76LL. This method is based on measuring the main parameters of the gas turbine engine operation (GT, n, Fc) and thermo-gas-dynamic parameters of the flow (P, P, T) in characteristic sections of the engine flow section and calculating the engine thrust using a number of correction factors (nozzle thrust coefficient ψc and flow coefficient air through the engine μvv), obtained during earth bench tests.

The aerodynamic method is based on measuring the longitudinal overload and mass of the aircraft during acceleration and braking at a given height with a constant number M of flight. This method is not independent, because to determine traction in braking modes, the gas-dynamic method is usually used, which can be applied in unforced mode of engine operation. This method is effective in determining thrust on a single-engine aircraft.

The dynamometric method consists in measuring the force transmitted to the aircraft pylon at the engine mounts. In this case, the calibration of the strain gauge suspension in the engine layout on a flying laboratory is a complex issue. Direct measurement of the thrust of a gas turbine engine in flight provides several advantages, since the measurement results are not affected by the flow conditions at the engine inlet and in the jet nozzle.

All of the above methods require special preparation of the engine, which cannot be implemented on a standard power plant in operation.

Monitoring the state of the aircraft engine that is in operation is very important. There are quite a few power equipment monitoring systems that use multi-sensor systems. In [1, 2], methods and a system for diagnosing heat power equipment that have multifunctional use are displayed. Also, indirect methods for calculating parameters are used for diagnosis, which are applicable for monitoring the parameters of on-board systems [3, 4]. In some cases, methods of virtual maintenance and basic ergonomics are applicable [5].

To control the parameters of aircraft structural elements, an analysis of the stress-strain state is used [6, 7]. Detection of unstable operating modes of structural elements is carried out using vibrations [8, 9] A method for changing the concentration of oxygen in air was presented in [10, 11], which can also be applied for measuring engine thrust to eliminate a methodological error.

The developed information-analytical systems that are implemented as laboratories on aircrafts [12, 13] are of considerable interest. In some cases, the test bench may be suspended on springs to an aircraft engine [14]. In [15] a system is described that implements the measurement of the stress-strain state of pylon structures based on strain gauge sensors, but the authors do not use it to measure the thrust of an aircraft engine.

2 Methods and Means

2.1 Method for Measuring of Aircraft Engine Thrust

The stages of the proposed method for determining traction are as follows:

- graduation of the pylon structure in the static test room; obtaining the load of the strain dependences ε_P at tensometric points on the load P applied to the vertical pin of the front node of the engine mount on the pylon (through this pin the engine thrust is transmitted to the aircraft structure) [16];
- calculation of strains ε_P by the finite element method (FE method) at the tensometric points of the pylon structure during a load P is applied to the pin of the front engine mount assembly;
- calculation of strains ε_T by the FE method at the tensometric points of the pylon structure during a load T is applied along the axis of symmetry of the geometric layout of the engine;
- determination of the thrust translation coefficient η as the ratio of deformations at the tensometric points of pylon structure during applying loads along the engine axis T and to the pin of the front linkage assembly P ($P = T$);
- measurement the thrust during the engines are running on the aircraft system; determination of structural deformation at tensometric points; taking into account the calibration curve and thrust translation coefficient η, it can be find the engine thrust.

2.2 Experimental Stand

To carry out the control calibration of strain gauge points [17] in the power structure of the pylon, an experimental stand was used, which was developed and mounted in the statistical test room of the Strength laboratory of the M. E. Zhukovsky National Aerospace University "Kharkiv Aviation Institute". The laboratory may conduct an examination of the structural strength of aircraft in accordance with the laws of Ukraine. The stand includes the IL-76 aircraft pylon, fastening and structural loading systems, and a strain gauge system. The layout of the test bench is shown in Fig. 1.

The technical condition of the IL-76 pylon during transmission to the laboratory is shown in Fig. 2. It should be noted the absence of a large number of fasteners for the pylon skin panels, which was subsequently eliminated by the laboratory staff. As with the initial calibration, the pylon in the stand is installed upside down. Sealing at

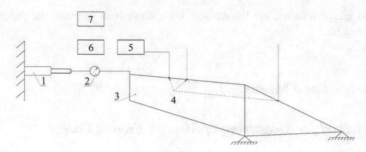

Fig. 1 Scheme of the experimental stand: **1**—hydraulic cylinder; **2**—dynamometer; **3**—plane's pylon; **4**—strain gauge points; **5**—measuring transducer; **6**—power supply with interface converter; **7**—computer

Fig. 2 Technical condition of the pylon

the attachment points of the pylon to the wing is carried out to the power floor using transition nodes.

For measuring the strain, KF 5P1-3-B-12 strain gages were used, the principle of operation of which is based on the property of the grating conductor to change the ohmic resistance due to tension or compression [18].

One-component cold hardening adhesive (cyacrine) recommended by the manufacturer of strain gages was used to label the strain gauges on metals. The surface of the structure was previously cleaned in the places where the sensors were sticking from dust and dirt, and with acetone and alcohol, from fats, oils and paints.

For measuring deformations in structural elements, strain gauges connected by a bridge circuit are used. This allows you to increase the sensitivity of the measured signal, to exclude temperature errors and the influence of unexplored load components.

The deformation of strain gauges assembled according to the bridge circuit is associated with the current load by the ratio [19]:

$$P = k \cdot \varepsilon, \tag{1}$$

where ε is the deformation arising from the action of the load P; k is the coefficient determined by graduation.

The sensors are connected by a bridge circuit and connected to a strain gauge system, which, after amplification and digitization, transmits a strain signal to a computer. The strain gauge system consists of a measuring transducer, a power supply with an interface transducer, a digital trunk cable, a null modem cable for communication with a computer, and a software package.

The measuring transducer is installed in a protective housing, which is stationary located in the pylon on the strut beam along with other standard equipment [20]. The converter allows to simultaneously poll 4 tensometric bridges.

The power supply together with the interface converter will be located in the cockpit and connected to a personal computer for processing, visualization and saving of measurement results.

3 Refinement of Places of the Sticker of Strain Gauge Sensors

In the work "Researches of the stressed-deformed state of the power structures of the plane", authors Fomichev, Zarutsky, Lovin, submitted for publication in the Aviation journal, the most rational places were chosen for sticking tensometric sensors to the power elements of the pylon. These places correspond to the pylon construction zones, in which there is an unambiguous dependence of the longitudinal deformations on the applied load, simulating the engine thrust. The calculations were carried out by the FE method in a geometrically and physically linear formulation.

According to the results of the calculations, it was found that the most rational places for the strain gauge stickers are the belts of the lower beam between 6 and 17 frames.

During the trial operation of the stand for measuring the thrust of the D30-KP engine on the basis of a non-flying aircraft of the IL-76 type, it was found that various tensometric points give, in terms of, slightly different engine thrust indicators. In the laboratory bench, the graduation of strain gauge sensors mounted on the pylon was carried out, which became one of the objectives of this article.

Table 1 Coordinates of the position of strain gauge points

Strain point number	Coordinate along the length of the belt from the pin, mm
1 (old)	1245
2 (old)	1840
3, 4 (old)	1470
5, 7 (new)	1170
6, 8 (new)	1050

It should be noted that the numerical and experimental data are in good agreement

The ambient temperature of a running engine increases significantly. The readings of strain gauge sensors at various air temperatures were verified. As a result, it was found that with uniform heating of all tensometric points soldered along the bridge circuit, a thermal compensation system works. However, when one point is locally heated, its readings, in terms of engine thrust, begin to differ from the readings of other points [21].

It was decided to choose new sticker locations for strain gauge sensors in order to eliminate the local effects of increased ambient temperature when the engine is running. Localization of sticker locations of strain gauge sensors is also possible by statistical methods [22, 23].

Table 1 shows the position coordinates of the old and new tensometric points along the length of the belt. The reference point is the vertical pin of the front engine mount on the pylon.

4 Method for Determining Thrust Translation Coefficient

According to the method for determining engine thrust according to tensometric measurements, it is necessary to know the value of the coefficient η. This coefficient can be found as the ratio of deformation at the tensometric point when applying traction along the axis of the engine to deformation at the same point when applying traction to the pin:

$$\eta = \varepsilon_T / \varepsilon_P. \tag{2}$$

Using the calculation model in the finite element package, we determined the strains at tensometric points under different conditions of application of maximum traction and reverse (Figs. 3 and 4).

The values of the coefficient η in the entire measured range of engine thrust for all strain gauge points are shown in Table 2.

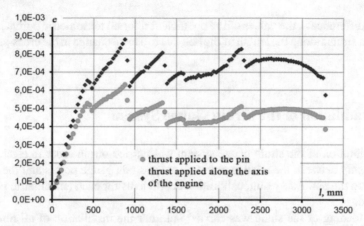

Fig. 3 Distribution of longitudinal deformation along the length of the belt of the lower beam for different applications of maximum traction

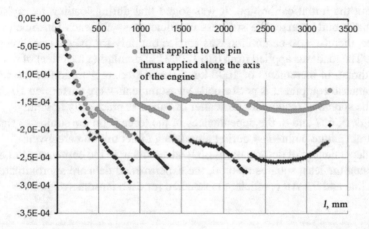

Fig. 4 Distribution of longitudinal deformation along the length of the belt of the lower beam of the pylon for different applications of reverse thrust

Table 2 Coefficients η for selected strain gauge points

Strain point number	η	Coordinate along the length of the belt from the pin, mm
1	1.500	1245
2	1.620	1840
3, 4	1.550	1470
5, 7	1.476	1170
6, 8	1.448	1050

The difference in the values of the coefficient η for all tensometric points is due to different strains along the length of the belt due to a local change in the cross-sectional area.

5 Graduation of the Strain Gauge System

The calibration of the strain gauge system was carried out in order to establish the relationship between the relative strain εP at the strain gauge points and the force P applied to the pin. As a result, calibration coefficients for each tensometric point are obtained.

The loading of the stand was carried out after the installation of all removable panels and hatches, as well as paneling. Strain gauge equipment located on the pylon was connected through a connector on the side panel in the area of the strut beam.

During the initial calibration, it was found that during loading the pylon with direct thrust and reverse, the stiffness coefficients A_P for each tensometric point are close, therefore, re-calibration was carried out only for the case of direct thrust loading. The load was applied from 0 to 12000 kg, simulating the effect of maximum engine thrust, in increments of 1000 kg. At each step, registration of the readings of tensometric equipment is performed. Measurements were performed 10 times to assess the spread of readings. The scatter of values is practically absent.

In Figs. 5, 6, 7 and 8, the dependences of the force P on the strains ε_P that arise at the strain gauge points 1–4 during simulating direct traction are given.

The dependences of deformations on load are linear in the entire range of applied load. Using the least squares method, the experimental data are approximated by a straight line and the AP coefficient is selected for each tensometric point.

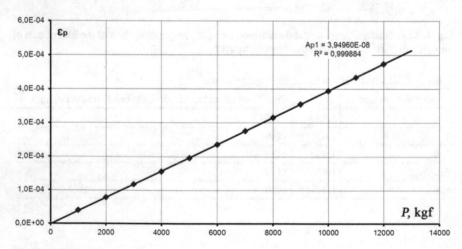

Fig. 5 Dependence of the force P on the strain at the strain gauge point 1

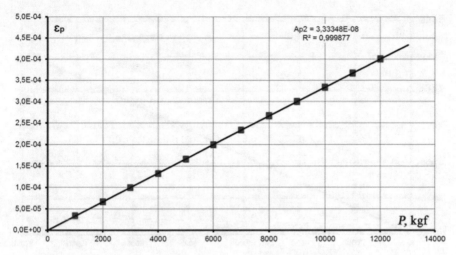

Fig. 6 Dependence of the force P on the strain at the strain gauge point 2

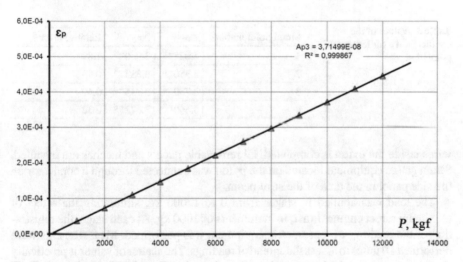

Fig. 7 Dependence of the force P on the strain at the strain gauge point 3

The results of the re-calibration data processing and comparison with the primary calibration data are shown in Table 3.

As can be seen from Table 3, the difference in the readings of the strain gauge sensors in terms of the coefficient A_P did not practically change during the year of operation of the traction measurement bench. This confirms the stability of the strain gauge system.

The calibration of new strain gauge points 5–8 was carried out. Previously, in each of them, strain gauge sensors are glued and soldered over a bridge circuit. A protective coating is applied to the sensors, installation and fastening of the switching

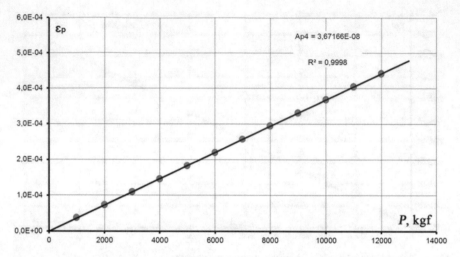

Fig. 8 Dependence of the force P on the strain at the strain gauge point 4

Table 3 Values of the coefficient A_P for tensometric points 1–4

Strain point number	A_{P1}	A_{P2}	Relative error ε, %
1	4.025^{-8}	3.950^{-8}	1.87
2	3.350^{-8}	3.333^{-8}	0.49
3	3.730^{-8}	3.715^{-8}	0.40
4	3.660^{-8}	3.672^{-8}	0.32

wires inside the pylon is completed, all removable panels and hatches are installed. Strain gauge equipment located on the pylon was connected through a connector on the side panel in the area of the strut beam.

The load was applied in stages from 0 to 12000 kg, simulating the effect of maximum direct engine thrust, in increments of 1000 kg. At each stage, the registration of the readings of tensometric equipment was performed. Measurements were performed 10 times to assess the spread of readings. The scatter of values is practically absent.

Figures 9, 10, 11 and 12 show the dependences of the force P on the strains ε_P that arise at the strain gauge points 5–8 during simulating direct traction.

All the dependencies in Figs. 9, 10, 11 and 12 are linear. The least squares method is used to approximate experimental data. The value of the coefficient A_P is obtained at each tensometric point. The approximation results are presented in Table 4.

The difference in the values of the A_P coefficient is associated with different sizes of deformation along the length of the belt due to local changes in the cross-sectional area and the individual properties of the sensors of the strain gauge bridge during the sticker process. This difference will not affect the final result of traction measurement, since each bridge (tensometric point) has its own calibration coefficient.

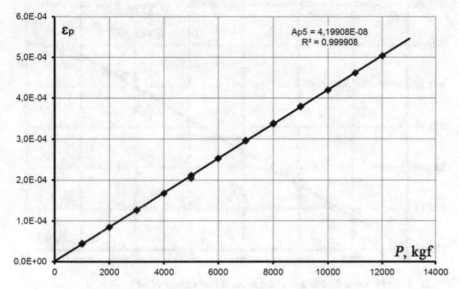

Fig. 9 Dependence of the force P on the strain at strain gauge point 5

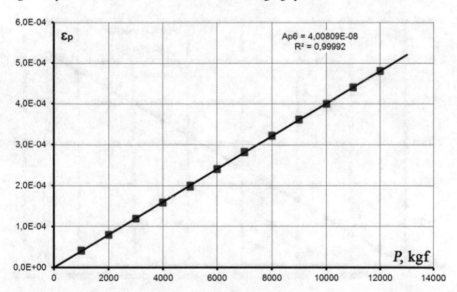

Fig. 10 Dependence of the force P on the strain at strain gauge point 6

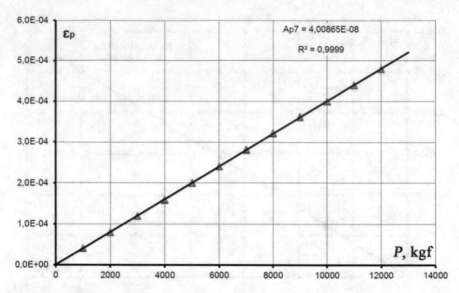

Fig. 11 Dependence of the force *P* on the strain at strain gauge point 7

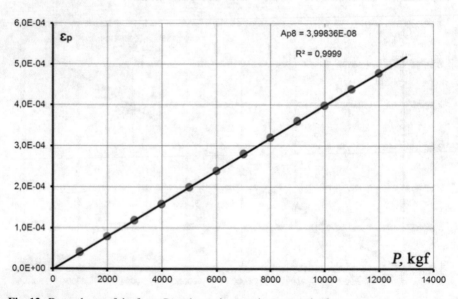

Fig. 12 Dependence of the force *P* on the strain at strain gauge point 8

6 Graduation of the Strain Gauge System

The adopted scheme of the sticker of the strain gauge sensors on the pylon belts allows to measure the longitudinal strain arising from the engine thrust. At the same time, placing the sensors included in the strain gauge point in one place and their

Table 4 Values of the coefficient A_P for tensometric points 5–8

Strain point number	A_P	Coordinate along the length of the belt from the pin, mm	Coefficient of determination R^2
5	$4.19908 \cdot 10^{-8}$	1170	0.999908
6	$4.00809 \cdot 10^{-8}$	1050	0.999920
7	$4.00865 \cdot 10^{-8}$	1170	0.999900
8	$3.99836 \cdot 10^{-8}$	1050	0.999900

wiring diagram make it possible to exclude the influence of temperature changes from engine operation. However, this approach does not allow directly taking into account the effect on the magnitude of the deformation of the reactive moment of the engine, which is transmitted to the lateral rods of the front mount.

Since the pylon structure and the loads acting on it are symmetrical with respect to the longitudinal axis, under the action of the reactive moment, the deformations of one belt will increase, and the second, respectively, decrease by the same amount. To eliminate this phenomenon during determining engine thrust in tests, it is necessary to take measurements simultaneously at tensometric points located on both pylon belts in the same cross section (the sensors were taken into account during labeling the sensors). After that, the obtained thrust values must be averaged.

The validity of the proposed methodology for taking into account the reactive moment during measuring the engine thrust is carried out using the FE model of the pylon. The accuracy of the calculation model was assessed by comparing the design deformations in the lower pylon belts with those measured by tensometry. The comparison results are shown in Fig. 13. It should be noted that the agreement between the calculated and measured strains is quite satisfactory.

Figure 14 shows the dependence of the change in deformation along the length of the pylon belts for two design cases:

- simulation of traction 12,000 kg without taking into account the action of the reactive moment;
- simulation of traction 12,000 kg, taking into account the action of the reactive moment.

To exclude the action of the reactive moment, the arithmetic mean value of the deformations of the left and right pylon belts is calculated. Figure 15 shows a comparison of belt average strain and strain calculated without torsion.

A theoretical analysis of the deformation of the pylon belts confirms the complete coincidence of the calculation results without taking into account the reactive moment and the method of accounting for the reactive moment.

Thus, in order to exclude the influence of the engine reactive moment in the tests, the thrust must be fixed at tensometric points on both pylon belts in the same cross section and its arithmetic average value must be calculated.

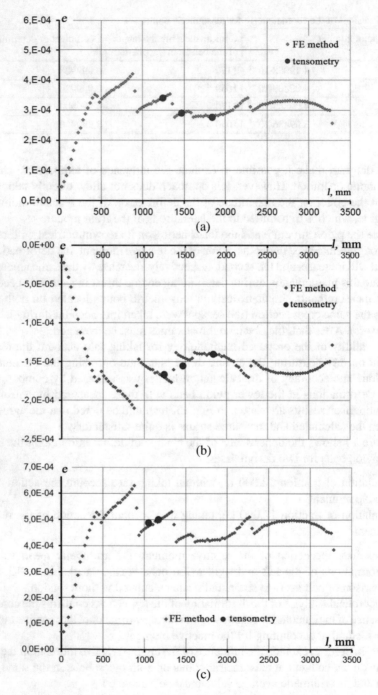

Fig. 13 Comparison of the calculation results with the strain gauge data of the pylon structure: (**a**) direct thrust of 8,000 kg; (**b**) reverse thrust of 4,000 kg; (**c**) direct draft 12,000 kg

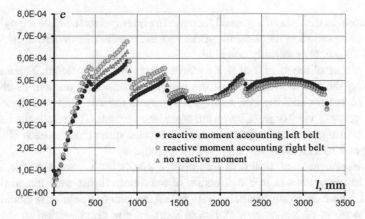

Fig. 14 Dependence of the change in deformation along the length of the pylon belts

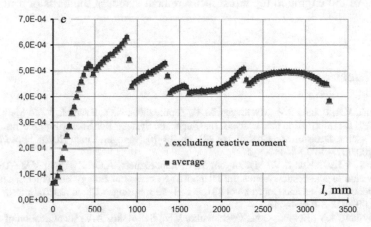

Fig. 15 Comparison of belt average strain and deformation calculated without torsion

7 Conclusions

The stand for the calibration of the strain gauge points of the pylon was mounted. Metrological verification of the measuring dinanometer and pressure testing of the hydraulic loading system is performed.

Re-calibration of strain gauge points 1–4, previously installed on the pylon carried out. The difference in calibration readings carried out with an interval of 1 year for the first bridge does not exceed 1.9%, for the rest—no more than 0.5%. This confirms the stability of the strain gauge equipment and its installation technology.

The effect of increasing ambient temperature on the readings of strain gauge sensors and, as a result, the estimated value of the engine thrust is checked. It has been established that local heating introduces an error in the calculation of engine thrust for a specific tensometric point relative to other points with a normal ambient temperature. In the case of a general increase in air temperature inside the pylon, the thermal compensation mechanism works and the results of calculating the engine thrust at all points practically coincide.

New places of the sticker of strain gauge points 5–8 are selected. The results of calculating the stress-strain state of the pylon structure with the aim of choosing rational places for sticking strain gauge sensors and the availability of these places from the point of view of installation are taken into account. To exclude the influence of local heating, new strain gauge points are located quite close to each other (at a distance of 120 mm along the length of the pylon).

The dependences of strains at tensometric points on the value of the force P applied to the pylon pin in the direction of maximum thrust were experimentally obtained. The calibration coefficients A_P for all tensometric points are calculated.

The location of tensometric points in pairs in one cross section of the pylon is justified. This approach allows you to take into account the influence of the jet moment of the engine in the thrust measurement stand on the basis of a flightless aircraft.

References

1. Babak, V.P., Babak, S.V., Myslovych, M.V., Zaporozhets, A.O., Zvaritch, V.M.: Methods and models for information data analysis. In: Diagnostic Systems For Energy Equipments. Studies in Systems, Decision and Control, vol 281, pp. 23–70. Springer, Cham (2020). https://doi.org/10.1007/978-3-030-44443-3_2

2. Babak, V.P., Babak, S.V., Myslovych, M.V., Zaporozhets, A.O., Zvaritch, V.M.: Technical provision of diagnostic systems. In: Diagnostic Systems For Energy Equipments. Studies in Systems, Decision and Control, vol 281, pp. 91–133. Springer, Cham (2020). https://doi.org/10.1007/978-3-030-44443-3_4

3. Kanyshev, A.V., Korsun, O.N., Ovcharenko, V.N., Stulovskii, A.V.: Identification of aerodynamic coefficients of longitudinal movement and error estimates for onboard measurements supercritical angles of attack. J. Comput. Sys. Sci. Int. **57**(3), 374–389 (2018). https://doi.org/10.1134/S1064230718030048

4. Babak, S., Babak, V., Zaporozhets, A., Sverdlova, A.: Method of statistical spline functions for solving problems of data approximation and prediction of objects state. In: CEUR Workshop Proceedings, vol. 2353, pp. 810–821 (2019). http://ceur-ws.org/Vol-2353/paper64.pdf

5. Xu, H.C., Pan, Y., Xiong, Y.X., He, J. W.: Comparative study on two civil engine thrust reversers for their maintainability analysis. In: MATEC Web of Conferences, vol. 169, p. 01026. EDP Sciences (2018). https://doi.org/10.1051/matecconf/201816901026

6. Fomichev, P.A.: Method for the evaluation of the service life under random loading based on the energy criterion of fatigue fracture. Strength Mater. **40**(2), 224–235 (2008). https://doi.org/10.1007/s11223-008-9011-5

7. Fomichev, P.A.: Substantiation of the predicted fatigue curve for structural elements of aluminum alloy. Strength Mater. **43**(4), 363 (2011). https://doi.org/10.1007/s11223-011-9305-x

8. Babak, V., Filonenko, S., Kalita, V.: Acoustic emission under temperature tests of materials. Aviation **9**(4), 24–28 (2008). https://doi.org/10.3846/16487788.2005.9635914

9. Babak, V., Filonenko, S., Kornienko-Miftakhova, I., Ponomarenko, A.: Optimization of signal features under object's dynamic test. Aviation **12**(1), 10–17 (2008). https://doi.org/10.3846/1648-7788.2008.12.10-17

10. Zaporozhets, A.: Analysis of control system of fuel combustion in boilers with oxygen sensor. Periodica Polytech. Mech. Eng. **64**(4), 241–248 (2019). https://doi.org/10.3311/PPme.12572

11. Babak, V.P., Babak, S.V., Myslovych, M.V., Zaporozhets, A.O., Zvaritch, V.M.: Principles of construction of systems for diagnosing the energy equipment. In: Diagnostic Systems For Energy Equipments. Studies in Systems, Decision and Control, vol. 281, pp. 1–22. Springer, Cham (2020). https://doi.org/10.1007/978-3-030-44443-3_1

12. Mirabbashi, A.S., Mazidi, A., Jalili, M.M.: Analytical and experimental flutter analysis of a typical wing section carrying a flexibly mounted unbalanced engine. Int. J. Struct. Stab. Dyn. **19**(02), 1950013 (2019). https://doi.org/10.1142/S0219455419500135

13. Bauer, M., Friedrichs, J., Wulff, D., Werner-Spatz, C.: Development and validation of an on-wing engine thrust measurement system. In: ASME Turbo Expo 2017: Turbomachinery Technical Conference and Exposition. American Society of Mechanical Engineers Digital Collection (2017, June). https://doi.org/10.1115/GT2017–63277

14. Zhu, H., Du, F., Zhao, S., Xu, Z., Lv, Z.: Design of a new type thrust measuring system for micro-turbojet engine. In: 2016 2nd International Conference on Artificial Intelligence and Industrial Engineering (AIIE 2016). Atlantis Press (2016, November). https://doi.org/10.2991/aiie-16.2016.109

15. Bauer, M., Friedrichs, J., Wulff, D., Werner-Spatz, C.: Measurement quality assessment of an on-wing engine thrust measurement system. In: ASME Turbo Expo 2018: Turbomachinery Technical Conference and Exposition. American Society of Mechanical Engineers Digital Collection (2018, June). https://doi.org/10.1115/GT2018-76496

16. Skorupka, Z., Sobieszek, A.: Strain gauge pin based force measurement. J. KONES **25**(2), 335–340. Online: http://yadda.icm.edu.pl/baztech/element/bwmeta1.element.baztech-c058114d-aa82-4b7e-8bfd-9f85eec80e8e

17. Stepanova, L., Petrova, E., Chernova, V.: Strength tests of a CFRP spar using methods of acoustic emission and tensometry. Russ. J. Nondest. Test. **54**(4), 243–248 (2018). https://doi.org/10.1134/S1061830918040101

18. Roskowicz, M., Leszczynski, P.: Evaluation of the suitability of the strain-gauge method for measuring deformations during the fatigue tests of aviation composite structures. Fatigue of Aircraft Struct. (9), 75–84 (2017). https://www.doi.org/10.1515/fas-2017-0006

19. Bol'shakova, A., Volobuev, V., Gorbushin, A., Petronevich, V.: Examination of systematic bench errors for calibration of a strain-gauge balance. Measur. Techn. **60**(8), 763–770. https://doi.org/10.1007/s11018-017-1268-2

20. Makhutov, N.: Generalized regularities of deformation and fracture process. Herald Russ. Acad. Sci. **87**(3), 217–228 (2017). https://doi.org/10.1134/S1019331617030030

21. Norbert, A., Huminic, A., Antonya, C.: Flight control system design and analysis of a light sport aircraft with emphasis on multibody dynamics and aerodynamic analysis. INCAS Bull. **10**(2), 221–229 (2018). http://doi.org/10.13111/2066-8201.2018.10.2.20

22. Eremenko, V., Zaporozhets, A., Isaenko, V., Babikova, K.: Application of wavelet transform for determining diagnostic signs. In: CEUR Workshop Proceedings, vol. 2387, pp. 202–214 (2019). Online: http://ceur-ws.org/Vol-2387/20190202.pdf

23. Zaporozhets, A., Eremenko, V., Isaenko, V., Babikova, K.: Approach for creating reference signals for detecting defects in diagnosing of composite materials. In: Shakhovska, N., Medykovskyy, M.O. (eds.) Advances in Intelligent Systems and Computing IV, vol. 1080, pp. 154–172. Springer, Cham (2020). https://doi.org/10.1007/978-3-030-33695-0_12

Development of a Virtual Scientific and Educational Center for Personnel Advanced Training in the Energy Sector of Ukraine

Yulii Kutsan(iD)**, Viktor Gurieiev**(iD)**, Andrii Iatsyshyn**(iD)**, Anna Iatsyshyn**(iD)**, and Evgen Lysenko**(iD)

Abstract The article substantiates the importance of using modern web-oriented technologies advanced training of the energy sector experts, in particular, the operational and dispatching personnel. As a result, the conducted research allowed substantiating the academic framework, structure, and functions of the virtual scientific and educational centre for the personnel professional development in the energy sector of Ukraine, including the knowledge assessment, training, and development of the personnel key competences. The advantages of using the distributed environment for the organization of operational personnel education and training using the simulation tools for the operating modes of electric energy systems in the virtual center are considered. An important practical aspect of the designed distance learning courses is the software application for the continuous development of key competencies of the energy sector personnel for rapid recognition of emergencies and, if necessary, their rapid elimination. This will allow them to acquire knowledge and practical skills to solve the problems of analysis, simulation, forecasting, and visualization of the operation modes monitoring data of large electric energy systems. Therefore, to improve the qualification of the personnel in the energy sector of Ukraine, a virtual scientific and educational center was designed, developed and implemented, and its scientific assistance was provided. Besides, a full-featured web mode simulator was developed and implemented. We believe that the use of a virtual scientific and educational center for the advanced training of energy industry specialists is an important component of the digitalization of the society and the energy sector of Ukraine.

Y. Kutsan · V. Gurieiev · A. Iatsyshyn (✉) · E. Lysenko
Pukhov Institute for Modelling in Energy Engineering of NAS of Ukraine, Kyiv, Ukraine
e-mail: iatsyshyn.andriy@gmail.com

V. Gurieiev
e-mail: viktor.gurieiev@ipme.com.ua

A. Iatsyshyn · A. Iatsyshyn
State Institution "Institute of Environmental Geochemistry" of NAS of Ukraine, Kyiv, Ukraine

Institute of Information Technologies and Learning Tools, NAES of Ukraine, Kyiv, Ukraine

© The Editor(s) (if applicable) and The Author(s), under exclusive license
to Springer Nature Switzerland AG 2020
V. Babak et al. (eds.), *Systems, Decision and Control in Energy I*, Studies in Systems,
Decision and Control 298, https://doi.org/10.1007/978-3-030-48583-2_5

Keywords ICT · Energy sector · Simulators · Operational and dispatching personnel · Virtual scientific and educational center · Advanced training

1 Introduction

The biggest challenge within the context of sustainable development of modern civilization today is the reliable operation of electric energy systems generating and distributing power energy for consumers in the right quality and quantity. The United Energy System (UES) of Ukraine integrates a large number of technological equipment distributed throughout the country for the generation, storage, transportation, distribution and use of power energy. The power-generating equipment must be controlled by highly qualified personnel. The inadequate personnel skill level and lack of preparedness to quickly eliminate the incidents lead to emergencies and enormous material costs to restore energy supply [1, 2].

The crisis in education, training and personnel development in the power industry exists not only in Ukraine but also all over the world. The main unresolved problems of the operational and dispatching personnel training system in the energy sector of Ukraine are the following: unadjusted legislative support; facilities and equipment are outdated; the system requires updating and development of the academic framework; the greater half of the personnel is close to retirement age or retired, therefore, younger employees are required.

The abovementioned problems cannot be solved effectively only by creating new educational institutions, training additional highly qualified teaching staff, and developing new curricula and training courses. Affordable and equal opportunities should be created for the continuous training of personnel of all electric utilities in the industry.

These problems, common for the EU and Ukraine, can only be solved by creating an internet-based unified global corporate education system, which integrates all existing educational institutions, virtual centres, training centres and staff training into a single unified system with a backbone managing and responsible organization.

In the current context, the purpose of operational and dispatching personnel training in the energy sector and the development of key competencies is:

(1) upgrading of the personnel skills to develop the key competences in order to recognize the conditions of cyber threats, emergencies and methods of emergency response, which involves the application of theoretical and practical methods related to the processes of generation and distribution of energy within the interdisciplinary context [3];
(2) training of highly qualified and professional dispatching personnel, the "top leaders of the industry", capable of solving scientific problems and issues regarding the optimization of the modes of operation of large EES, as well as to solve practical problems of the digital transformation of the industry [4].

Moreover, power-engineering professionals must be skilled enough to use information and communication technologies (ICT) to help model and predict the conditions of system emergencies in power systems, therefore the ability to apply these technologies is important for future professional activity. Considering the fact of constant improvement of ICT and the development of new management systems, it is important to introduce the latest developments, systems, software to the staff, as well as to develop the skills to use these tools in their future professional activities.

2 Literature Analysis and Problem Statement

The analysed scientific publications in the field of energy sector have been structures by the following areas of activity: development and diagnostics of power equipment [5–7]; cybersecurity of energy facilities [8–11]; theoretical methods and practical tools for mathematical and computer simulation in the energy sector [12–15]; development of mathematical and software tools for assessing the impact of fuel and energy enterprises on the economic component of the country and the environment [16–23]; training systems in the energy sectors based on virtual technologies [1, 3, 4] etc.

Various aspects of the development and use of virtual laboratories (centres) for educational purposes are considered in the works of foreign scientists: R. Heradio [24], L. de la Torre [24], D. Galan [24], F. Cabrerizo [24], J. Grodotzki [25], T. Ortelt [25], R. Morales-Menendez [26], R. A.Ramírez-Mendoza [26], M. Hernández-de-Menéndez [27], A. Vergara [28], M. Barker [29], S. Olabarriaga [29], M. Krasnjanskyj [30], B.Paljukh [31], V. Belov [31], I. Obrazcov [31] etc.; national researchers: H. Biletska [32], K. Bobrivnik [33], M. Gladka [33], M. Kiktev [33], O. Semenikhina [34], V. Shamonia [34], I. Stepura [35] etc.

According to the analysis of the scientific publications and authors' own professional experience, the main directions in research of the creation of virtual scientific-educational centers (VSEC) for the training of operational and dispatching personnel in the energy sector of Ukraine have been identified and generalized as follows:

- development of a new or improvement of the existing methodological base of VSEC, as well as the approaches to ensure the centres' scientific background;
- development of automated systems for monitoring of unauthorized impacts on critical infrastructural facilities;
- development and application of distance learning courses (DLCs) to organize the personnel blended learning process;
- development and use of conceptual, external and internal schemes of distributed power engineering enterprise database;
- development and use of mathematical methods and simulating tools for the operating modes of large EES;
- development of FWMS information support, including software development in the form of distributed computer systems and databases.

Some of the aspects outlined above have already been addressed in previous publications by the authors of this article, in particular [1, 3, 4]. Specific features of education and advanced training of the experts in the power-engineering sector and other areas have been investigated in [2, 33, 36–44]. However, the publications discussed above, pay insufficient attention to the use of integrated e-learning technologies for education and training of operational and dispatching personnel in the energy sector, the creation of an appropriate training and methodological base for the organization of a blended form of professional development of personnel using virtual distributed training systems.

3 Purpose and Objectives of the Study

The purpose of the research is to design a virtual scientific and educational centre for advanced training of specialists in the energy sector of Ukraine.

To achieve this purpose, the following tasks were set:

1. To substantiate the academic framework, structure, and functions of the virtual scientific and educational centre for the personnel professional development in the energy sector of Ukraine.
2. To develop and describe a full-featured web mode simulator.

4 Research Methods

A number of methods were used in this study: the analysis, systematization, study of practical experience in solving problems of VSEC; the method of topological analysis of large UES; the selection of specialized application and system software for solving the problems of VSEC organization and functioning; the simulation of man-made load on the UES of Ukraine; the analysis of training experience of future energy and environmental experts; the method of structuring theoretical materials for the development of academic framework for staff education and training.

5 Research Results

The digitalization of society requires higher education institutions and research institutions to consistently develop and implement modern ICT trends, demonstrate the ability to solve digital transformation problems, enabling them to significantly increase their competitiveness, attract additional resources, and improve academically. Important trends in education are: the use of cloud technologies, access to virtual computer systems, business analytics and Internet of things, virtual laboratories, augmented reality (AR) and virtual reality (VR) technologies etc. [45, 46].

We agree that a wide range of different physical processes can be simulated and investigated using training laboratories. However, the existing facilities and resources of educational institutions cannot ensure personal access to such equipment for everyone. One of the possible ways to solve the problem is to use online laboratories in the educational process [35].

The concept of "online laboratory" means software-hardware complexes for conducting experiments without direct contact with physical equipment or in its absence [35]. Based on the principle of operation, the online training laboratories are divided into remote, virtual and hybrid ones. Remote laboratories include laboratory equipment and software for managing and digitizing the data obtained. A student has the opportunity to set the mode characteristics, control the appropriate mechanisms, take off data from control devices and record them for further processing [30]. Virtual laboratories are software complexes for conducting experiments without the use of a physical laboratory facility [31]. During the experiment, the laboratory equipment operation is simulated and all objects and processes are being computer-simulated as well. In hybrid laboratories, hardware (locally or remotely controlled) is synchronized and connected to virtual laboratories. Such a combination significantly improves the perception of the object under study and helps to form a holistic picture of the world with clear relationships between the parameters of engineering systems [30].

An example of a hybrid lab is GOLDi (Grid of Online Laboratory Devices Ilmenau). GOLDi is a software and hardware training complex for virtual and remote experiments developed in Germany. The features of this virtual lab allow testing algorithms in Assembler and C programming languages on the models of such physical systems as an unloading station, elevator, production conveyor and automated warehouse. The experiment runtime is accessed through a personal user account on the GOLDi lab site [36].

The advantages of using virtual laboratories in the educational process are [33]: individual training without interruption; division of laboratory work into modules; gradual study of technological processes; possibility of simultaneous use at the lectures, practical and laboratory classes; analysis of experimental data at the time of the experiment; modification and improvement, adjustments to the existing model.

Virtual and remote labs reduce the costs associated with conventional hands-on laboratories with the necessary equipment, space, and support staff. Moreover, they provide additional benefits such as distance learning support, improved accessibility of the laboratory for people with disabilities, and increased security for hazardous experiments [24].

The requirements for engineering education are changing along with increasing demand from the industry, promoting the transformation of the fourth industrial revolution. For this reason, the German Federal Ministry of Education and Research (BMBF) has launched a number of education-related research projects, the largest of which is the collaborative research project "ELLI—Excellence in Teaching and Learning in Engineering Science". Among the three universities involved, namely:

RWTH Aachen University, Ruhr-Universität Bochum and TU Dortmund University, the latter is at the centre of the development of remote and virtual laboratories for mechanical engineering education. The created Massive-Open-Online-Course (MOOC) includes remote laboratories as part of the instructional method used. In order to overcome the boundaries of remote laboratories, the virtual facilities of these laboratories and a common virtual experimental laboratory were developed. They are designed for a variety of devices that allow students to visually explore and experiment on complex processes. All these tools are being integrated into different disciplines. The range of the laboratories developed is being expanded to include new processes and the introduction of AR technologies. The AR devices will be used to develop new laboratories. These labs will be integrated into the existing labs to enhance the user experience and prepare for the future of mixed reality, in which AR information processing will be an important skill [25].

Recent developments in ICT, the Internet of things and systems digitalization have made it possible to develop simulators for manufacturing plants and production systems with high-definition images and realism in their animation, which can perfectly simulate real systems. The integration of software with real automation equipment/devices opens up new ways to improve and accelerate the teaching/learning process and the development of competencies and skills. A hybrid lab combines virtual and remote lab approaches to teach industrial automation courses [26].

Virtual laboratories and virtual research environments are the terms used to refer to digital, community-created environments that are designed to meet the needs of the research community. In particular, they imply integrated access to research community resources, including software, data, collaboration tools, workflows, tools and high-performance computing, typically via the Internet and mobile applications. Scientific gateways, virtual laboratories, and virtual research environments make significant contributions to many research fields, promoting more effective, open and reproducible research in new ways [29].

Thus, the use of virtual laboratories in the educational process allows, on the one hand, to get practical skills for conducting experiments, to get acquainted in detail with a virtual prototype of modern equipment, to investigate the processes and phenomena to be at risk for fire and explosion without worrying about the possible consequences. On the other hand, there is an opportunity to organize the interaction of the virtual laboratory complex with the actual operating equipment of different enterprises for the collection of experimental data, which will provide an appropriate level of scientific development and technological control of the process [33].

Ukrainian Sumy State University has developed its own methodological model of online courses, which promotes a high level of interactivity of educational content, in particular the widespread use of virtual training equipment and simulators. More than 2,000 pieces of virtual training equipment and simulators have been created based on Java, JS, Flash, Unity3D (including VR and AR). In 2019, the University established a VR and AR scientific and research laboratory, which is part of the online learning ecosystem and allows reaching a new level of VR/AR application in education in the future. The University's ecosystem development plan envisages the complete transition of the university's educational process to e-learning technology, the active introduction of blended learning models, the use of VR and AR for learning, the development of its own VR online course concept, the increase and the active extension of non-academic, online courses making it possible to transfer the results into academic disciplines [46, 47].

Having analysed the scientific literature and considering many years' experience in the field of power engineering, the main problems of the current system of training of operational and dispatching personnel of the fuel and energy complex enterprises, given in Table 1, are identified.

So far, the current system of training and upgrading of skills of the energy sector personnel of Ukraine is very inefficient, has significant problems and disadvantages. The existing network of higher education institutions is no longer optimal for quality training, knowledge control and the development of the key competences of the fuel and energy complex staff. Such condition of the personnel training system of the energy sector of Ukraine does not allow effectively solving the problems of work with the personnel, and therefore requires significant modernization in many aspects.

An innovative virtual environment (consisting of a full-featured web mode simulator (FWMS) and an online training course) for the personnel on-site and distance training of the Ukrainian NPPs was developed in close cooperation with leading institutes, higher education institutions and power engineering experts in order to partially solve the problems outlined above. Figure 1 shows the page of the author's website http://www.infotec.ua. There are several sections on this website, which present a virtual model of the UES of Ukraine, the articles regarding power industry, webinars, information on substation automation, distance learning and training of operational and dispatching staff of NPC "Ukrenergo", etc. Figure 2 shows some of the developed DLCs for the skills improvement of the energy sector of Ukraine personnel.

The purpose of the DLC (The Technology of Routine Switching in Electric Power System of the UES of Ukraine) is to prepare future specialists for the development of their knowledge and key competencies in recognizing the emergency conditions, the ability to timely prevent and quickly eliminate their consequences, including the ability to use specialized training simulation systems. The substantive modules of the DLC include "Expertise in Recognition of Emergency Conditions", "Expertise in Fast Methods of Accidents Elimination", "Expertise in Electrical Engineering and Power Industry Fundamentals", "Methods of Optimization of EES Operation".

The DLCs consist of lectures and practical classes, it is planned to hold consultations and independent out-of-class work of operational staff to study additional and scientific literature. 60 h/2 ECTS credits are allocated for studying the discipline (16 h of lectures, 24 h of practical work, 20 h of independent work).

The main tasks of the DC are:

- to introduce the legal support in the field of the routine switching technology in the electric power systems to the staff;
- to teach how to evaluate the operational state of the electric power systems in different regions of Ukraine;

Table 1 Problems of the current system of training of operational and dispatching personnel of the fuel and energy complex enterprises

Levels	Problematics
Legislative	Lack of national standards for vocational education of personnel for particular power industry professions
	Unadjusted legislative support
	Lack of industry's professional standards for the training system of operational and dispatching personnel
Technological	Facilities and resources are outdated
	Lack of open information and simulating environment for carrying out proper studies, analysis and forecasting of normal and emergency modes of operation of the EES and UES, including all existing levels of the hierarchy of management as a whole, in order to use the results of research and calculations in the training system
	The existing full-featured mode simulators operating at nuclear power plants (NPPs) of Ukraine are the most effective means of simulation training with the exception of emergency electromagnetic transients in the EES and UES. At the same time, they are very expensive and typically focused on simulation of particular NPP power-generating machinery, which complicates the transfer and application of the acquired skills and knowledge to eliminate accidents in other similar parts of large EES, and does not allow to simulate the emergency consequences of power modes of EES or UES equipment for the NPPs operating modes
	A small number of task-specific simulators
Educational	Requires updating and development of the academic framework
	A large number of the personnel professions in the electric-power industry operating a variety of power-generating machinery, and the lack of its unification, significantly limits the development of a common approach (common standards of advanced training) to the creation of modern and effective personnel training systems for electricity industry as a whole
	The advanced training programs applied in the relevant educational institutions are very outdated and tend to focus on specific theoretical or practical often irrelevant issues related to the operation of EES and UES equipment and are based on electrical engineering guidelines rather than information models of EES and UES in general
	A limited number of open web-based shared resources to obtain competency-based structured knowledge, including advanced training for teachers and training specialists with the obligatory control of their qualifications
Personnel	No proper real-time and long-term psychophysiological expertise (diagnostics) of the personnel: testing and system of vocational and social rehabilitation
	There are no scientifically substantiated criteria for determining the level of reliability of key competences of the staff (by means of annual simple testing), which provides the basic technological processes of generation, distribution and consumption of electric and thermal power

(continued)

Table 1 (continued)

Levels	Problematics
	Lack of proficiency testing of teaching staff providing professional training for electric-power industry personnel
	The problem of the loss of experience, skills and knowledge and practical experience of highly qualified staff, including those who are retiring. It is possible to solve this problem through the remote involvement of retired specialists with extensive experience into the development of scenarios of emergency training and training exercises, participation in distance courses, expert discussion of possible conditions of predicted accidents (or the accidents already happened)
	The greater half of the personnel is close to retirement age or retired, therefore, younger employees are required

Fig. 1 Webpage of the author's site http://www.infotec.ua

- to teach how to predict the operation status of power equipment of the UES of Ukraine using mathematical simulation techniques;
- to introduce the basic concepts and methods used in assessing the impact of critical infrastructure operational personnel;
- to develop skills to solve the problems of VSEC applying special methods and means;
- to develop the ability to use specialized software-simulating systems in the field of power engineering;
- to develop independent work skills.

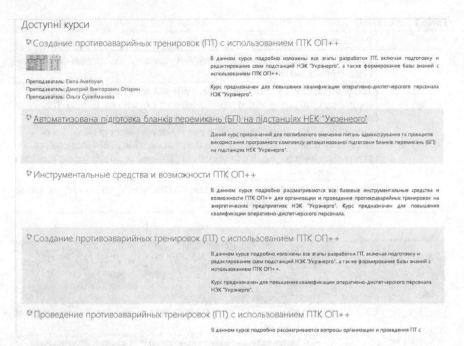

Fig. 2 The developed DLCs to improve the personnel skills in the energy sector of Ukraine on the website http://www.infotec.ua

During the lecture course, the main issues concerning the technology of routine switching in electric networks and dispatch control are presented. During the workshops, it is proposed to develop scenarios of complex accidents that have occurred or are foreseen, as well as to create emergency response drill (ERD) based on them. The created ERDs are numbered (without surnames) and subsequently used during the workshops only for the purpose of their constructive discussion and improvement. During the discussion sessions, the staff discusses all the proposed scenarios of the ERD scenarios, identifies their shortcomings and forms an assessment. The feasibility of using this approach is manifested in the ability to identify the best software developers and significantly expand the possibility of designing the software within the context of possible accidents.

The software developed by the authors is aimed at solving many scientific and practical problems of VSEC. The main ones are [1, 3, 4]:

- collection, storage and processing of operational data of current modes of EES operation;
- sampling of mode monitoring data, their graphical visualization and statistical analysis;
- design and definition of actual problems of VSEC;
- determination of the predicted scenario of accidents and design of the corresponding ERDs;

- study of static and dynamic stability of the EES (study of the risk dynamics over different periods of time);
- determination of the mutual influence of different operation modes of NPP units of the UES as a whole;
- visualization of EES operating modes using graphs, diagrams, electronic maps etc.

The pages of the site http://www.infotec.ua describe in detail the stages of the educational process using FWMS. The DLC operates on the Moodle e-learning system combined with FWMS, which is designed to develop and maintain stable skills of the personnel of the Ukrainian UES (all levels of the existing hierarchy of management) to quickly eliminate the conditions of occurrence and possible development of major system emergencies. The opportunity to participate in emergency response drills (ERD) is presented to the personnel of nuclear, thermal, wind and solar power plants, hydroelectric power plants and power companies (Oblenergo).

The distributed system of databases and servers of applications of the data-processing network (DPN) FWMS is integrated into a global network and connected to the Internet. Separate databases and application servers, some of which can be virtual, are combined, like the power plants of the EES, into a single global computer network and are geographically located in different places (all types of power plants, power companies, etc.). Such DPN FWMS structure allows to adequately and quickly simulate various emergencies or voltage and frequency modes of parallel operating power systems and/or interconnections [48].

The simulation results are easily accessible to any user, providing effective organization and delivery of inter system DPNs and training exercises to staff who have access to the Internet and are located in any DPN-suitable workplaces or training centres (TCs). These may be the territories of NPPs, TPPs, HPSs, high-voltage substations, power companies and other energy enterprises. Depending on the type of DPN scenario, the features of the emergencies and the objectives of the DPNs, it is possible to form and involve any staff of operational and control personnel of the enterprises of the UES of Ukraine to eliminate the specific emergency. At the same time, there are no reasons, limitations and necessity to conduct DPN in one place [48].

It also provides an opportunity, if necessary, to involve any staff of enterprises of Russia, Moldova etc. operating in parallel with the UESs. The main conditions for the possibility of staff participation in the DPN are the availability of a suitable password and the Internet access from a PC, located at their own workplace or TC. The start time and duration of the DPN are set (chosen from the menu) by the instructor. He can also set the start time of the DPN (any time of the day within 24 h), the discretization of the load/generation change (minimum is 30 s), and the duration of the DPN for the following possible options: increase, decrease, minimum or maximum load/generation of the ES. The Chief Emergency Training Director/Instructor and the participants (local instructor and trainee) should have different access rights to the simulator resources. All members and guests are required to enter a login and password to enter the FWMS. Guest (minimum) access only allows to observe and view the diagrams [48].

Figures 3 and 4 present FWMS operation examples.

To date, the developed software, which is a component of the virtual scientific-educational centre, was implemented at the National Technical University of Ukraine "Igor Sikorsky Kyiv Polytechnic Institute", a private joint-stock company "National

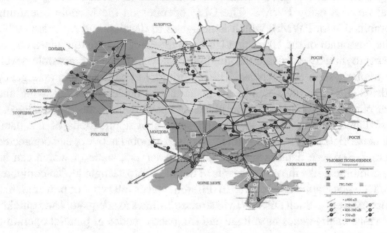

Fig. 3 Example of FWMS operation

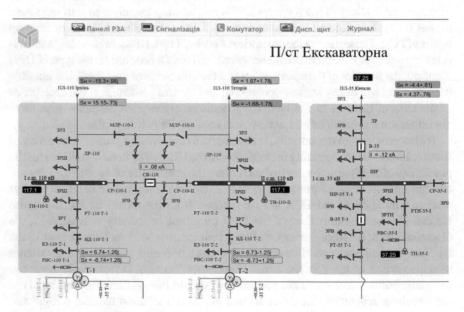

Fig. 4 Example of FWMS operation

Energy Company Ukrenergo" and others. Over 5,000 people have already upgraded their professional skills using the virtual scientific and educational centre.

6 Conclusions

The current state of the HR system of Ukraine is characterized by significant short-comings in the context of personnel education and virtual training in the energy sector of Ukraine, which does not allow the fulfilment of obligations to international organizations in full.

In order to improve the qualification of the personnel in the energy sector of Ukraine, a virtual scientific-educational centre was developed and its scientific support was provided; a DLC for the organization of blended staff training as well as FWMS were developed and implemented.

As a result of the conducted research, the academic, structure and functions of the virtual scientific-educational centre for the professional development of personnel in the energy sector of Ukraine are substantiated, including control of knowledge, training and development of key competences. An important practical aspect of the authors' DLCs is the application of the software for the development of the key competencies for energy industry staff to rapidly recognize emergencies and, if necessary, eliminate them. This will allow them to acquire knowledge and practical skills to solve the problems of analysis, simulation, forecasting and visualization of monitoring data of the modes of operation of large EES.

Therefore, an application of FWMS will help:

- ensure the possibility of continuous introduction of the routine switching technology in the electrical networks to the operating personnel and the fundamentals of the dispatcher operation control;
- introduce the basic concepts and methods used in assessing the state of EES objects and the mutual influence of their modes of operation to the personnel;
- training of personnel to solve current problems of professional production activity with the use of web-oriented training systems and special methods and tools for simulation the modes of operation of the EES, as well as developing the ability to quickly recognize the emergency conditions and their elimination.

References

1. Gurieiev, V., Sanginova, O., Avetisyan, E.: Modeling methods and means of construction and operation of virtual research and training centers in energy. Monograph. VP « Edel'veys » , Kiev, (2019)
2. Popov, O., Iatsyshyn, A., Kovach, V., Artemchuk, V., Taraduda, D., Sobyna, V., Sokolov, D., Dement, M., Yatsyshyn, T., Matvieieva, I.: Analysis of possible causes of NPP emergencies

to minimize risk of their occurrence. Nuc. Radia. Saf. **1**(81), 75–80 (2019). https://doi.org/10.32918/nrs.2019.1(81).13

3. Gurieiev, V., Sanginova, O.: Simulation and study of modes for full-scale mode simulator for Ukrainian energy systems. In: Proceedings of the 2nd International Conference on Intelligent Energy and Power Systems (IEPS'2016), Kyiv, Ukraine June 7–11, IEEE (2016). https://doi.org/10.1109/ieps.2016.7521848

4. Gurieiev, V., Sanginova, O.: Distributed simulation environment of modes for full-scale mode simulator for Ukrainian energy systems. Tech. Electrodyn. **5**, 67–69 (2016). https://doi.org/10.15407/techned2016.05.067

5. Zaporozhets A.: Development of software for fuel combustion control system based on frequency regulator. In: CEUR Workshop Proceedings, vol. 2387, pp. 223–230 (2019). Online: http://ceur-ws.org/Vol-2387/20190223.pdf

6. Zaporozhets, A.: Analysis of control system of fuel combustion in boilers with oxygen sensor. Period. Polytech. Mech. Eng. **63**(4), 241–248 (2019). https://doi.org/10.3311/PPme.12572

7. Zaporozhets, A., Kovtun, S., Dekusha, O.: System for monitoring the technical state of heating networks based on UAVs. In: Shakhovska, N., Medykovskyy, M. (eds.) Advances in Intelligent Systems and Computing IV. CCSIT 2019. Advances in Intelligent Systems and Computing, vol. 1080. Springer, Cham (2020). https://doi.org/10.1007/978-3-030-33695-0_61

8. Mokhor, V., Gonchar, S., Dybach, O.: Methods for the total risk assessment of cybersecurity of critical infrastructure facilities. Nucl. Rad. Saf. **2**(82), 4–8 (2019). https://doi.org/10.32918/nrs.2019.2(82).01

9. Mokhor, V.V., Tsurkan, O.V., Tsurkan, V.V., Herasymov, R.P.: Information security assessment of computer systems by socio-engineering approach. In: Selected Papers of the XVII International Scientific and Practical Conference on Information Technologies and Security (ITS 2017), Kyiv, Ukraine, November 30, 2017, pp. 92–98 (2017). http://ceur-ws.org/Vol-2067/paper13.pdf

10. Kichak, V.M., Rudyk, V.D., Gonchar, S.F.: Compensation of non-stationary temporal errors of the measurement channel. Telecommun. Radio Eng. **69**(10), 869–880 (2010). https://doi.org/10.1615/TelecomRadEng.v69.i10.30

11. Vysotska, O., Davydenko, A.: Keystroke pattern authentication of computer systems users as one of the steps of multifactor authentication. In: Hu, Z., Petoukhov, S., Dychka, I., He, M. (eds.) Advances in Computer Science for Engineering and Education II. ICCSEEA 2019. Advances in Intelligent Systems and Computing, vol. 938. Springer, Cham (2020). https://doi.org/10.1007/978-3-030-16621-2_33

12. Saukh, O., Papst, F., Saukh, S.: Synchronization games in P2P energy trading. In 2018 IEEE International Conference on Communications, Control, and Computing Technologies for Smart Grids, SmartGridComm 2018. Institute of Electrical and Electronics Engineers Inc. (2018). https://doi.org/10.1109/smartgridcomm.2018.8587421

13. Chemeris, A., Lazorenko, D., Sushko, S.: Influence of software optimization on energy consumption of embedded systems. In: Kharchenko, V., Kondratenko, Y., Kacprzyk, J. (eds.) Green IT Engineering: Components, Networks and Systems Implementation, pp. 111–133. Springer, Cham (2017). https://doi.org/10.1007/978-3-319-55595-9_6

14. Vynnychuk, S., Misko, V.: Acceleration analysis of the quadratic sieve method based on the online matrix solving. East. Eur. J. Enter. Technol. **2**(4–92), 33–38 (2018). https://doi.org/10.15587/1729-4061.2018.127596

15. Vynnychuk, S., Maksymenko, Y., Romanenko, V.: Application of the basic module's foundation for factorization of big numbers by the Fermat method. East. Eur. J. Enter. Technol. **6**(4–96), 14–23 (2018). https://doi.org/10.15587/1729-4061.2018.15087

16. Kyrylenko, O.V., Blinov, I.V., Parus, Y.V., Ivanov, H.A.: Simulation model of day ahead market with implicit consideration of power systems network constraints. Techn. Electrodyn. **5**, 60–67 (2019). https://doi.org/10.15407/techned2019.05.060

17. Blinov, I.V., Parus, Ye.V., Ivanov, H.A.: Imitation modeling of the balancing electricity market functioning taking into account system constraints on the parameters of the IPS of Ukraine mode. Techn.l Electrodyn. **6**, 72–79 (2017). https://doi.org/10.15407/techned2017.06.072

18. Popov, O., Iatsyshyn, A., Kovach, V., Artemchuk, V., Taraduda, D., Sobyna, V., Sokolov, D., Dement, M., Yatsyshyn, T.: Conceptual approaches for development of informational and analytical expert system for assessing the NPP impact on the environment. Nucl. Radia. Saf. **3**(79), 56–65 (2018). https://doi.org/10.32918/nrs.2018.3(79).09

19. Popov, O., Iatsyshyn, A., Kovach, V., Artemchuk, V., Taraduda, D., Sobyna, V., Sokolov, D., Dement, M., Hurkovskyi, V., Nikolaiev, K., Yatsyshyn, T., Dimitriieva, D.: Physical features of pollutants spread in the air during the emergency at NPPs. Nucl. Radia. Saf. **4**(84), 88–98 (2019). https://doi.org/10.32918/nrs.2019.4(84).11

20. Bogorad, V., Bielov, Y., Kyrylenko, Y., Lytvynska, T., Poludnenko, V., Slepchenko, O.: Forecast of the consequences of a fire in the chernobyl exclusion zone: a combination of the hardware of the mobile laboratory RanidSONNI and computer technologies DSS RODOS. Nucl. Radia. Saf. **3**(79), 10–15 (2018). https://doi.org/10.32918/NRS.2018.3(79).02

21. Kovach, V., Lysychenko, G.: Toxic Soil Contamination and Its Mitigation in Ukraine. In: Dent, D., Dmytruk, Y. (eds.) Soil Science Working for a Living. Springer, Cham (2017). https://doi.org/10.1007/978-3-319-45417-7_18

22. Mergner, R., Janssen, R., Kovach, V., et al.: Fostering sustainable feedstock production for advanced biofuels on underutilised land in Europe. Eur. Biomass Conf. Exhib. Proc. **2017**, 125–130 (2017)

23. Yatsyshyn, T., Shkitsa, L., Popov, O., Liakh, M.: Development of mathematical models of gas leakage and its propagation in atmospheric air at an emergency gas well gushing. Eastern-European Journal of Enterprise Technologies 5/10(101), 49–59 (2019). https://doi.org/10.15587/1729-4061.2019.179097

24. Heradio, R., Torre, L.D., Galan, D., Cabrerizo, F.J., Dormido, S.: Virtual and remote labs in education: A bibliometric analysis. Comput. Edu. **98**, 14–38 (2016). https://doi.org/10.1016/j.compedu.2016.03.010

25. Grodotzki, J., Ortelt, T.R., Tekkaya, A.E.: Remote and virtual labs for engineering education 4.0. Proc. Manufac. **26**, 1349–1360 (2018). https://doi.org/10.1016/j.promfg.2018.07.126

26. Morales-Menendez, R., Ramírez-Mendoza, R.A., Guevara, A.V.: Virtual/remote labs for automation teaching: a cost effective approach. IFAC-PapersOnLine **9**(52), 266–271 (2019). https://doi.org/10.1016/j.ifacol.2019.08.219

27. Hernández-de-Menéndez, M., Guevara, A.V., Morales-Menendez, R.: Virtual reality laboratories: a review of experiences. International Journal on Interactive Design and Manufacturing, 1–20. (2019). https://doi.org/10.1007/s12008-019-00558-7

28. Vergara, A., Rubio, M.P., Lorenzo, M.: On the design of virtual reality learning environments in engineering. Multi. Technol. Int. **1**, 11 (2017)

29. Barker, M., et al.: The global impact of science gateways, virtual research environments and virtual laboratories. Future Gen. Comput. Sys. **55**, 240–248 (2019). https://doi.org/10.1016/j.future.2018.12.026

30. Krasnjanskyj, M.N.: Development of school virtual laboratories on the basis of programming environment LabVIEW, (2017). http://clubedu.tambov.ru/methodic/2007/virt

31. Paljukh, B.V., Belov, V.V., Obrazcov, I.V.: The technology of virtual laboratories in the practice of building education. Build. Mater. Equip. Technol. XXI Century **1**, 42–45 (2013)

32. Biletska, G.A.: Usage of virtual laboratory works while training professional ecologists ». Inf. Technol. Edu. **12**, 44–49 (2012). https://doi.org/10.14308/ite000314

33. Bobrivnyk, K., Gladka, N., Kiktev, M.: Designing virtual learning laboratory for students of technical and technological professions. Energy Autom. **3**, 18–23 (2014)

34. Semenikhina, O.V., Shamonia, V.H.: Virtual laboratories as instrument of educational and scientific activity. Teach. Sci. Theory Hist. Innovative Technol. **1**(11), 341–345 (2011)

35. Stepura, I.S.: Possibilities for the use of GOLDi hybrid laboratory for educational experiments. Open Educational E-Environment of Modern University 3, (2017). https://doi.org/10.28925/2414-0325.2017.3.33036

36. Manual Control of the "3-Axis-Portal". http://goldi-labs.net/index.php?Site=37

37. Iatsyshyn, Anna V., Kovach, V.O., Romanenko, Ye.O., Iatsyshyn, Andrii V.: Cloud services application ways for preparation of future PhD. In: Kiv, A.E., Soloviev, V.N. (eds.) Proceedings

of the 6th Workshop on Cloud Technologies in Education (CTE 2018), Kryvyi Rih, Ukraine, December 21, 2018, CEUR Workshop Proceedings, vol. 2433, pp. 197–216 (2018). http://ceur-ws.org/Vol-2433/paper12.pdf

38. Shkitsa, L.Y., Panchuk, V.G., Kornuta, V.A.: Innovative methods of popularizing technical education. Proceedings of the Conference Innovative Ideas in Science 2016, Baia Mare, Romania, November 10–11, 2016, IOP Conference Series: Materials Science and Engineering, vol. 200, p. 012023 (2016). https://doi.org/10.1088/1757-899x/200/1/012023

39. Zuban, Yu., Lavryk, T., Ivanets, S.: Integrated development environment for distant courses based on project approach. Techn. Sci. Technol. **4**(6), (2016)

40. Mintii, I.S., Shokaliuk, S.V., Vakaliuk, T.A., Mintii, M.M., Soloviev, V.N.: Import test questions into Moodle LMS. In: Kiv, A.E., Soloviev, V.N. (eds.) Proceedings of the 6th Workshop on Cloud Technologies in Education (CTE 2018), Kryvyi Rih, Ukraine, December 21, 2018, pp. 529–540 (2018). http://ceur-ws.org/Vol-2433/paper36.pdf

41. Markova, O.M., Semerikov, S.O., Striuk, A.M., Shalatska, H.M., Nechypurenko, P.P., Tron, V.V.: Implementation of cloud service models in training of future information technology specialists. In: Kiv, A.E., Soloviev, V.N. (eds.) Proceedings of the 6th Workshop on Cloud Technologies in Education (CTE 2018), Kryvyi Rih, Ukraine, December 21, 2018, pp. 499–515 (2018). http://ceur-ws.org/Vol-2433/paper34.pdf

42. Radionova, L.V., Chernyshev, A.D., Lisovskiy, R.A.: Interactive educational system—virtual simulator "sheet rolling". Proc. Eng. **206**, 512–518 (2017). https://doi.org/10.1016/j.proeng.2017.10.509

43. Pochtoviuk, S.I., Vakaliuk, T.A., Pikilnyak, A.V.: Possibilities of application of augmented reality in different branches of education. In: Kiv, A.E., Shyshkina, M.P. (eds.) Proceedings of the 2nd International Workshop on Augmented Reality in Education (AREdu 2019), Kryvyi Rih, Ukraine, March 22, 2019, CEUR Workshop Proceedings, vol. 2547, pp. 92–106 (2019). http://www.ceur-ws.org/Vol-2547/paper07.pdf

44. Vakaliuk, T.A., Kontsedailo, V.V., Antoniuk, D.S., Korotun, O.V., Mintii, I.S., Pikilnyak, A.V.: Using game simulator Software Inc in the Software Engineering education. In: Kiv, A.E., Shyshkina, M.P. (eds.) Proceedings of the 2nd International Workshop on Augmented Reality in Education (AREdu 2019), Kryvyi Rih, Ukraine, March 22, 2019, CEUR Workshop Proceedings, vol. 2547, pp. 66–80 (2019). http://www.ceur-ws.org/Vol-2547/paper05.pdf

45. Iatsyshyn, Anna V., Kovach, V.O., Romanenko, Ye.O., Deinega, I.I., Iatsyshyn, Andrii V., Popov, O.O, Kutsan, Yu.G., Artemchuk, V.O., Burov, O.Yu., Lytvynova, S.H.: Application of augmented reality technologies for preparation of specialists of new technological era. In: Kiv, A.E., Shyshkina, M.P. (eds.) Proceedings of the 2nd International Workshop on Augmented Reality in Education (AREdu 2019), Kryvyi Rih, Ukraine, March 22, 2019, CEUR Workshop Proceedings, vol. 2547, pp. 181–200 (2019). http://ceur-ws.org/Vol-2547/paper14.pdf

46. Iatsyshyn, Anna V., Kovach, V.O., Lyubchak, V.O., Zuban, Y.O., Piven, A.G., Sokolyuk, O.M., Iatsyshyn, Andrii V., Popov, O.O, Artemchuk, V.O.: Application of augmented reality technologies for education projects preparation. Proceedings of the 7th Workshop on Cloud Technologies in Education (CTE 2019), Kryvyi Rih, Ukraine, December 20, 2019, CEUR Workshop Proceedings (2019, in press)

47. Sumy State University Online Learning Ecosystem: Design for competition of XI International Exhibition "Innovation in Contemporary Education—2019". Sumy State University, Sumy (2019)

48. Full-featured Web Simulator PORT. http://www.infotec.ua/uk/node/82

Analysis of the Air Pollution Monitoring System in Ukraine

Artur Zaporozhets[iD], **Vitaliy Babak**[iD], **Volodymyr Isaienko**[iD],
and Kateryna Babikova[iD]

Abstract In this chapter, the basic features of the functioning of the monitoring system of air pollution in Ukraine are considered. The process of implementing Directive 2008/50/EU in Ukrainian legislation, the requirements for monitoring various pollutants in the air (particulate matter (PM2.5, PM10), CO, SO_2, NO_2, O_3) and the measures that Ukraine must take to implementation of the Directive are considered. The approaches to monitoring air pollution in Ukraine, the features of the Ukrainian air quality index—air pollution index are considered. The state of air pollution in different cities of Ukraine is analyzed, the most polluted cities and regions are determined in accordance with the air pollution index. Statistical studies on emissions into the air by stationary and mobile sources of pollution in Ukraine in 1990–2018 are carried out. In particular, the following pollutants were considered: SO_2, NO_2, CO, CO_2, PAHs, Zn, Pb, Cu, Cr, Ni, Ar. It has been established that 5 cities in Ukraine generate emissions, which make up 40.2% of all emissions in the country. The main enterprises polluting the air in the largest cities are given. The process of measuring air pollutant concentrations as part of an air pollution monitoring system is considered. The optimal number of posts required for the implementation of the Directive has been determined. The features of monitoring air pollution in Kyiv are considered. The above-mentioned shortcomings of the air pollution monitoring system are shown and ways to eliminate them are proposed. A modern information-analytical system for monitoring air pollution is proposed.

Keywords Air pollution · Air quality monitoring · Control system · Monitoring system · Ukraine · Directive · Pollutants · Stationary posts · Wireless sensors network

A. Zaporozhets (✉) · V. Babak
Institute of Engineering Thermophysics of NAS of Ukraine, Kyiv, Ukraine
e-mail: a.o.zaporozhets@nas.gov.ua

V. Isaienko · K. Babikova
National Aviation University, Kyiv, Ukraine

V. Babak et al. (eds.), *Systems, Decision and Control in Energy I*, Studies in Systems, Decision and Control 298, https://doi.org/10.1007/978-3-030-48583-2_6

1 Introduction

Given the trend of Ukraine's foreign policy on the prospects of joining the European Union (EU), the main direction of further development in the field of atmospheric air monitoring should be the consistent harmonization of the relevant elements of atmospheric air observation networks with EU requirement [1–3]. The signing of the Association Agreement between Ukraine and the EU and its member states has opened up new opportunities for the implementation of environmental standards.

For ensuring the implementation of the basic principles of the functioning of the state system of environmental monitoring, it is necessary to attract the existing potential of all monitoring entities, primarily on the basis of the coherence and progressiveness of the regulatory and methodological support of observation networks [4].

For Ukraine, the implementation of EU legislation in the field of environmental protection takes place within 8 sectors and is regulated by 29 sources of law—Directives and EU regulations, which establish general rules and standards that should be reflected in domestic law. Unlike modern environmental legislation of Ukraine, EU law sources determine the quantitative and qualitative results that each country needs to achieve within a certain period of time and determine the procedures that must be implemented to achieve these results. A feature of the EU Directives is that states must adapt their legislation to achieve the goals defined by the Directives, but at the same time they determine the methods for achieving them.

Monitoring of air quality is regulated by 6 directives:

- Directive 1999//32/EC on sulfur in liquid fuels.
- Directive 98/70/EC on quality of gasoline and diesel fuel.
- Directive 94/63/EC on control of volatile organic compounds (VOCs).
- Directive 2004/42/EC on paints and varnishes.
- Directive 2004/107/EC on As, Cd, Hg, Ni and Polycyclic Aromatic Hydrocarbons (PAHs) in ambient air.
- Directive 2008/50/EC on ambient air quality and cleaner air for Europe [5].

Directive 2008/50/EU, which defines the framework requirements for monitoring and evaluating the quality of atmospheric air and according to which Ukraine has to implement its individual provisions, deserves special attention [6, 7].

In particular, to establish zones and agglomerations throughout their territory according to the degree of air pollution, as well as the procedure for their revision. In cases where the levels of pollutants exceed any of the regulatory limits or there is a risk of such excess, there is a need to develop action plans for air quality for the respective territories. In Ukraine, such a classification has not been previously used, and the relevant plans were prepared exclusively by administrative territorial division.

This Directive also sets the basic limit values for protecting public health:

- for average annual PM10—40 $\mu g/m^3$, the 24-hour limit value is 50 $\mu g/m^3$, cannot be exceeded more than 35 times during a calendar year;
- for PM2.5, the target value and the limit value for stage 1—annual average—25 $\mu g/m^3$;
- for PM2.5, the limiting value for stage 2 is the annual average—20 $\mu g/m^3$;
- for SO2, the hourly limit value is 350 $\mu g/m^3$; it cannot be exceeded more than 24 times during a calendar year; 24-hour limit value—125 $\mu g/m^3$, cannot be exceeded more than 3 times during a calendar year;
- for average annual NO_2—40 $\mu g/m^3$, hourly limit value—200 $\mu g/m^3$, cannot exceed more than 18 times during a calendar year;
- for average annual lead—0.5 $\mu g/m^3$;
- for annual average benzene—5 $\mu g/m^3$;
- for CO, the limiting daily 8-hour value is 10 mg/m^3;
- for O_3, the target value is the maximum daily 8-hour value—120 $\mu g/m^3$, cannot be exceeded for more than 25 days during a calendar year for 3 years.

In addition to air quality standards, the Directive establishes:

- rules for assessing atmospheric air quality (upper and lower thresholds for assessing, measuring, modeling, combining, data quality objectives);
- principles for the preparation of local, regional or national plans for improving air quality, including a list of information to be included and short-term action plans, including their detailed content;
- principles for determining zones and agglomerations;
- reporting to the European Commission on air quality;
- requirements for the availability of information to the public.

Since the Directive was signed in 2015, by the end of 2020 Ukraine should implement the following tasks [8]:

- adopt national legislation about zones and identify authorized agencies;
- establish upper and lower bounds on the assessment of target and limit values for the main pollutants, and goals to reduce the impact of PM2.5;
- determine zones and agglomerations throughout Ukraine by the degree of air pollution, as well as the procedure for revising the classification of zones and agglomerations depending on assessment thresholds;
- establish public information systems;
- establish air quality assessment systems;
- introduce air quality plans for zones and agglomerations where the level of pollution exceeds the border/target value;
- establish short-term action plans for agglomerations where there is a risk of exceeding permissible pollution limits.

2 Approaches to Monitoring Air Pollution in Ukraine

Air pollution depends on the amount of air emissions, their specifics and weather conditions [9–11].

The level of air pollution is determined by comparison with the maximum permissible concentration (MPC), in particular their single and daily average values. The single and average daily MPC values of some pollutants, the concentrations of which are measured at stationary posts, are shown in Table 1.

The data in Table 1 indicate that benz(a)pyrene is the most hazardous, which refers to substances of hazard class 1. Other listed substances belong to hazard classes 2 and 3.

For a quantitative assessment of air pollution in individual settlements, an atmospheric pollution index (API) is used, which is defined as the sum of the ratios of actual concentrations of the five most important additives to their average daily MPC:

$$API = \sum_{i=1}^{n} \left(\frac{q_i}{MPC_{iMS}} \right)^{\alpha_i}, \tag{1}$$

where n is the amount of impurities taken into account in the calculation; q_i is the concentration of the i-th substance; MPC_{iMS}—the maximum single MPC of the i-th substance, α_i—the ratio of the hazard of the i-th substance to the substance of 3 hazard class ($\alpha_1 = 1.5$; $\alpha_2 = 1.3$; $\alpha_3 = 1$; $\alpha_4 = 0.85$).

According to the API value, the level of atmospheric air pollution is divided: (1) clean air (API < 2.5); (2) slightly polluted air ($2.5 \leq$ API < 7.5); (3) polluted air ($7.5 \leq$ API < 12.5); (4) highly polluted air ($12.5 \leq$ API < 22.5); (5) high polluted air ($22.5 \leq$ API < 52.5); (6) extremely polluted air (API $\geq$ 52.5).

Table 1 MPC of some pollutants in the air in Ukraine

№	Pollutants	Formula	Max single MPC, $\mu g/m^3$	Daily average MPC, $\mu g/m^3$	Hazard Class
1	Dust	–	0.5	0.15	3
2	Sulfur dioxide	SO_2	0.5	0.05	3
3	Carbon monoxide	CO	5	3	4
4	Nitrogen monoxide	NO	0.4	0.06	3
5	Nitrogen dioxide	NO_2	0.085	0.04	2
6	Ammonia	NH_3	0.2	0.04	4
7	Formaldehyde	CH_2O	0.035	0.003	2
8	Hydrogen chloride	HCl	0.2	0.2	2
9	Hydrogen fluoride	HF	0.02	0.005	3
10	Benzo(a)pyrene	$C_{20}H_{12}$	–	10^{-6}	1

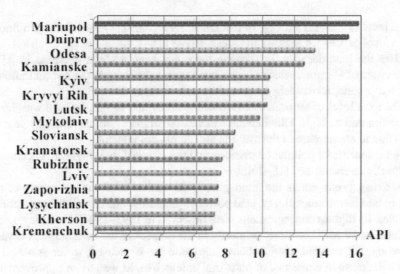

Fig. 1 The value of API in the most polluted cities of Ukraine in 2018

Assessment of air pollution in Ukrainian cities is carried out by the Borys Sreznevsky Central Observatory. In total, monitoring was carried out in 39 cities at 129 stationary posts of the monitoring network of hydrometeorological stations.

In the air 22 pollutants were determined, including 8 heavy metals [10]. The average annual concentration of formaldehyde in cities, where observation posts are present, was about $2.3\,MPC_{DA}$, nitrogen dioxide—$1.5\,MPC_{DA}$, phenol—$1.3\,MPC_{DA}$.

According to API in 2018, a very high level of air pollution was observed in Mariupol and the Dnipro; high—in Odesa, Kamianske, Kyiv, Kryvyi Rih, Lutsk, Mykolaiv, Sloviansk, Kramatorsk, Rubizhne, Lviv, Zaporizhia, Lysychansk, Kherson, Kremenchuk. The high level of air pollution in these cities is due to significant concentrations of phenol, nitrogen dioxide, formaldehyde, hydrogen fluoride, carbon monoxide and suspended solids (dust) [12–14] (Fig. 1).

Most cities with polluted and very polluted air are in the Donetsk region—3 cities, in the Dnipropetrovsk region—3 cities, in the Luhansk region—2 cities, in the Poltava region—1 city.

In 2018, 3 cases of high atmospheric air pollution with CO in the city of Obukhov with a maximum concentration of $8.8\,MPC_{DA}$ were also recorded. There were no cases of extremely high air pollution in 2018 in Ukraine.

High daily average concentrations of formaldehyde were recorded in Mariupol—$6\,MPC_{DA}$; Dnipro—$5\,MPC_{DA}$; Mykolaiv, Odesa and Kryvyi Rih—4.7–$3.7\,MPC_{DA}$; in 9 cities—at the level of $3\,MPC_{DA}$. The excess of average annual concentrations of NO_2 was recorded in Kyiv, Dnipro, Kherson—at the level of 3.3–$2.8\,MPC_{DA}$, in 4 cities—at the level of $2.3\,MPC_{DA}$. Excess concentrations of suspended solids at the level of 3.0–$2.3\,MPC_{DA}$ were recorded in Kryvyi Rih, Dnipro, Kamianske; phenol (C_6H_5OH) at the level of $2,3\,MPC_{DA}$—in Kramatorsk, Kamianske, Sloviansk; CO at the level of 1.8–1.3 MPC_{DA} in Rubizhne, Lysychansk, Odesa; hydrogen fluoride at the level of 1.6–1.4 MPC_{DA} in Odesa, Rivne [15–17].

Over the past decades, in Ukraine there has been a slight decrease in API in some cities of Ukraine, which is associated with a decrease in the concentrations of benzo(a)pyrene, formaldehyde, ammonia and nitrogen dioxide.

The total level of air pollution in Ukraine according to API in 2018 was 7.6 and was estimated as high. This indicator increased slightly compared to last year (7.2 API) due to an increase in the average annual phenol content.

Next, sources of pollutant emissions and the cities of their highest concentration are briefly characterized [18–20].

Benz(a)pyrene enters the atmosphere with emissions from enterprises of non-ferrous and ferrous metallurgy, heat power engineering, and also during operating of vehicles. Its highest concentrations were observed in 1992–1994, when they reached the level of 10 MPC or more, primarily in the centers of the metallurgical industry. Currently, the concentration of benzo(a)pyrene has decreased, which is associated with a decrease in emissions of industrial enterprises, as well as an improvement in the fuels quality. Still quite significant concentrations of benzo(a)pyrene are observed in Zaporizhia, Sloviansk (2.7 MPC), Dnipro and Ternopil (2.3 MPC).

Sulfur dioxide is a characteristic impurity contained in emissions from chemical, metallurgical and paper industries. Sulfur dioxide is the second pollutant after carbon dioxide. Sulfur dioxide emissions cause acid rain. The highest concentrations of this substance are observed in Odesa and Kyiv.

Dust is generated during fuel combustion, in production processes, and also during soil erosion. Its highest concentrations are observed in the eastern and southern regions of Ukraine.

Carbon monoxide is a characteristic impurity formed during the incomplete combustion of fossil fuels. As a rule, elevated concentrations of carbon monoxide are observed at TPPs, boiler houses, and metallurgical enterprises. Carbon monoxide is a compound that actively reacts with atmospheric constituents and contributes to the greenhouse effect. High concentrations of CO are observed in Rubizhne, Zaporizhia, Odesa, and Kyiv.

Nitrogen dioxide enters the atmosphere during the burning of fossil fuels, as well as in the production of nitrogen fertilizers, paints, and synthetic fabrics. The highest concentrations of nitrogen dioxide are typical for Sloviansk (3 MPC), Kyiv (2.5 MPC), Odesa, Dnipro, Bila Tserkva.

Nitrogen oxides. The following nitrogen oxides are distinguished depending on the degree of oxidation: NO, N_2O, N_2O_3, NO_2, N_2O_5. Oxides N_2O_3 and N_2O_5 are solids, others are gases. Natural sources of nitric oxide are such natural phenomena as volcanic eruptions and lightning strikes. Anthropogenic sources of nitrogen oxides in the atmosphere are the chemical industry, the production of explosives, fertilizers, nitric acid, bacterial decomposition of silage, etc. The largest amount of nitrogen oxides in the atmosphere comes from automobile transport. The dynamics of changes in the concentration of nitrogen oxides in the air during the day is associated with the intensity of movement of vehicles and solar radiation. So, during daylight hours, the concentration of nitrogen oxides in the air increases significantly due to the processes of photochemical oxidation of nitrogen. Nitrogen oxides are a dangerous pollutant due to its high toxicity and impact on atmospheric phenomena (acid rain, smog). During chemical processes in the atmosphere, nitrogen oxides lead to the destruction of the ozone layer.

Hydrogen sulfide and carbon disulfide are inferior to the atmosphere both separately and together with other sulfur compounds. The main emission sources are sugar, artificial fiber, coke and oil refining industries. In the atmosphere, during interacting with other pollutants, it is oxidized to sulfur dioxide. The highest concentrations of hydrogen sulfide pollution were recorded in Toretsk, Mariupol, Zaporizhia.

Hydrogen fluoride enters the atmosphere with emissions from non-ferrous metallurgy enterprises, mineral fertilizer factories, and construction industry enterprises. Exceeding the MPC is observed in Sloviansk (2.8 MPC), Zaporizhia and Odesa (1.8 MPC).

Fluorine compound. Sources of pollution are industrial enterprises for the production of aluminum, glass, steel, enamels, ceramics, phosphorus fertilizers. Substances containing fluorine enter the atmosphere as gaseous compounds— hydrogen fluoride or dust of calcium and sodium fluoride. These compounds have a toxic effect.

Chlorine compound. These substances enter the atmosphere from chemical plants producing pesticides, hydrochloric acid, soda, bleach, organic dyes, and hydrolysis alcohol. The toxicity of chlorine depends on both the type of compound and its concentration.

Ammonia is contained in emissions of chemical industry enterprises, in particular specializing in the production of mineral fertilizers. An increase in the concentration of this pollutant is characteristic for Cherkasy (3 MPC), Kamianske (1.8 MPC), Mariupol (1.5 MPC).

Phenol enters to the atmosphere with emissions from the iron and steel industry. An increase in the concentration level is typical for Odesa (2 MPC), Kamianske, Toretsk, Zaporizhia (1.7 MPC) and Sloviansk (1.3 MPC).

3 Analysis of Air Pollution in Ukraine

According to the State Statistics Service of Ukraine, in 2018 the volume of pollutant emissions was 3866.7 thousand tons (for comparison, in 2017—3974.1 thousand tons, in 2016—4498.1 thousand tons), while emissions from stationary sources in 2018 were 2508.3 thousand tons (for comparison, in 2017—2584.9 thousand tons, in 2016—3078.1 thousand tons). The carbon dioxide emissions that have the greatest impact on the greenhouse effect in 2018 were 126 mln tones., which is 2% more than in 2017. At the same time, since 2008 in Ukraine there has been a clear tendency to reduce emissions of pollutants and carbon dioxide into the air (Figs. 2 and 3).

Figures 4, 5, 6 and 7 also provide data on the emissions of various pollutants into the air during 1990–2018.

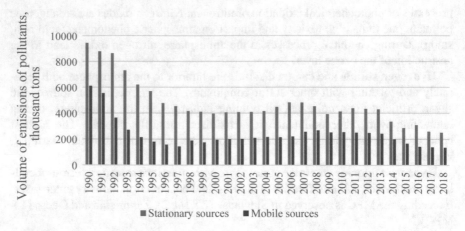

Fig. 2 The volume of emissions of pollutants into the air in the period 1990–2018

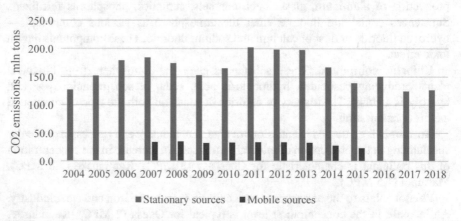

Fig. 3 Carbon dioxide emissions into the air during the period 2004–2018

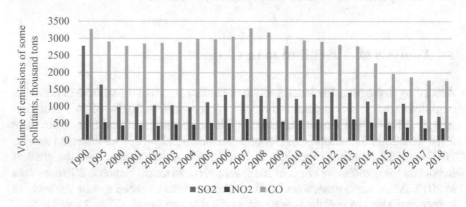

Fig. 4 Emissions of SO2, NO2, CO in Ukraine during 1990–2018

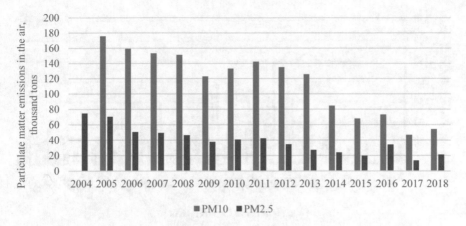

Fig. 5 PM10 and PM2.5 emissions in Ukraine during 2004–2018

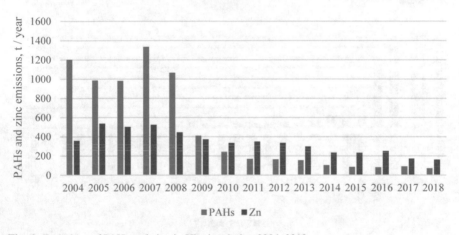

Fig. 6 Emissions of PAHs and zinc in Ukraine during 2004–2018

According to [1], the leaders among regions in terms of emissions of pollutants into the air from stationary sources of pollution are the Dnipropetrovsk and Donetsk regions (Fig. 8). In particular, in 2017, with total 2584.9 thousand tons of emissions, 657.3 and 784.8 thousand tons were emitted by the above regions.

Compared to the previous year, an increase in atmospheric emissions was recorded in 10 regions: Kyiv (68.6%), Kherson (29.1%), Odesa (26.4%), Zhytomyr (25.5%), Zakarpattia (24%), Cherkasy (19.8%), Ivano-Frankivsk (11.6%), Khmelnitsky (4.8%), Sumy (2.1%), Donetsk (0.7%).

Table 2 shows 10 regions which had the greatest impact on air pollution in 2018.

In general, for every resident of Ukraine in 2018, 59.3 kg of emissions of pollutants into the atmosphere fell. By territorial aspect, for every square kilometer of the country's territory, 4.35 tons of pollutants were accounted for. Among the settlements

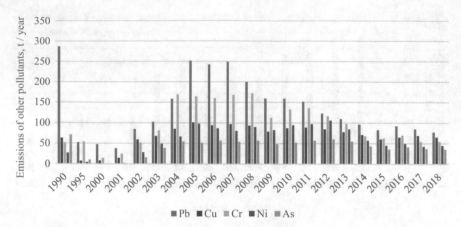

Fig. 7 Emissions of other pollutants (plumbum, copper, chromium, nickel, arsen) in Ukraine during 1990–2018

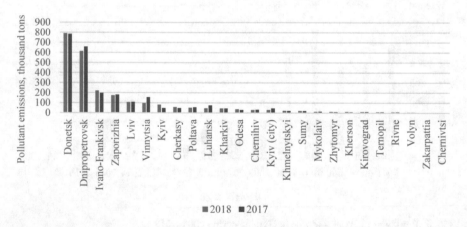

Fig. 8 The volume of pollutant emissions from stationary sources by region in 2017–2018

of the largest anthropogenic load (over 100 thousand tons of pollutants) 5 cities of Ukraine received, which are shown in Table 3. The total volume of pollutant emissions into the air of these cities from all emissions of the country is 40.2%.

During 2018, about 11 thousand industrial enterprises emitted pollutants into the air. From them, 2508.3 thousand tons of pollutants entered to the atmosphere, which is 3% less than in 2017.

The main pollutants of atmospheric air in Ukraine remain the enterprises of the processing and mining industries, the supply of electricity, gas and air conditioning, the total emissions of which consist for more than 92% of the total emissions from stationary sources.

Table 2 The most polluting regions of Ukraine in 2018

Region	Volume of emissions, thousand tons	% of total country emissions
Donetsk	790.2	31.5
Dnipropetrovsk	614.3	24.5
Ivano-Frankivsk	221.4	8.8
Zaporizhia	174.7	7.0
Lviv	106.7	4.3
Vinnytsia	97.3	3.9
Kyiv	81.3	3.2
Cherkasy	57.9	2.3
Poltava	52.1	2.1
Luhansk	46.7	1.9

Table 3 The dynamics of emissions of pollutants into the air in Ukrainian cities, thousand tons

City	2010	2011	2012	2013	2014	2015	2016	2017	2018	% from total in 2018
Mariupol	364.3	382.4	330.4	333.8	289.4	249.6	257.3	288.2	316.6	12.60
Kryvyi Rih	395.0	358.6	354.6	351.8	327.4	327.0	342.9	323.9	267.4	10.7
Burshtyn	146.8	198.7	174.7	182.7	199.8	198.0	168.5	160.1	182.9	7.3
Kurakhove	123.9	166.2	148.4	166.0	125.0	112.7	126.4	154.7	139.2	5.5
Kamianske	108.5	124.7	116.4	115.5	105.0	101.0	90.5	57.8	103.3	4.1

Among the emissions from economic activities, the largest share of pollutant emissions (excluding carbon dioxide emissions) is accounted for by the supply of electricity, gas and air conditioning—39.4%.

The second largest amount of pollutant emissions into the atmosphere is the processing industry. It accounts for 35.2%. At the same time, the share of metallurgical production is 29%. Mining and quarrying accounted for 17.7% of total air emissions.

4   Emissions in Major Cities of Ukraine

Kyiv. According to the State Statistics Service of Ukraine, on December 1, 2019, 2.966 million people live in Kiev. The area of the city is about 890 km^2.

The total volume of emissions into the atmosphere of the city, according to the Main Department of Statistics in Kyiv, has sharply decreased in recent years: in

2014—214.2 thousand tons; 2015—171.0 thousand tons; 2016—34.3 thousand tons; 2017—45.5 thousand tons; 2018—29.2 thousand tons.

The main source of air pollution in Kyiv is road transport. It accounts for about 83% of emissions. A lower level of pollution occurs at thermal power plants (TPPs), enterprises of machine-building, chemical, chemical and pharmaceutical industries, light, food industry and construction industry. Among the main emission sources, the following can be distinguished: CJSC Ukr-Kan-Power (TPP-4), TPP-5, TPP-6, Energia plant, OAO Ukrplastic, OAO Korchevatsky plant of building materials and structures, Darnitsky Car Repair Plant (DVRZ) [21–23].

The most polluted substance of the air in Kiev is nitrogen dioxide. The average annual concentration of this substance in recent years is about 2.5 MPC. Significant are also the concentrations of formaldehyde (3–1.3 MPC), benz(a)pyrene (1–1.2 MPC), phenol (1–1.3 MPC).

Significant differences in the location and specifics of industrial enterprises and highways lead to the fact that the total level of air pollution and its specificity in individual areas vary significantly.

So, atmospheric air near observation points located at the intersection of Victory Avenue and ul. Academician Tupolev is very polluted. The proximity of the observation post on the highway allows to determine that the atmospheric air in this section is very polluted with benz(a)pyrene, nitrogen dioxide and carbon monoxide. Also close to this level of pollution is the area within the Bessarabian square. Among the pollutants, carbon monoxide predominates here. The main reason for the pollution of this section is a large number of cars, significant traffic jams in front of traffic lights.

In general, the most polluted air in the central part of the city, the cleanest air within the forest park areas, especially in the western part of the city.

In recent years, the level of air pollution in the city remains stable. In comparison with the beginning of the 90 s, it decreased significantly due to the reduction in the number of industrial enterprises, reduction of emissions by enterprises, improvement of the quality of fuel materials, construction of new transport interchanges [24–27].

Dnipro. The population is about 1 million people. The main industries that affect the level of air pollution in the city are engineering, ferrous metallurgy, and electric power. The companies most affected by air quality are: Dnipro Metallurgical Plant, PJSC Dnipro Pipe Plant, Prydniprovska TPP, OJSC INTERPIPE NTZ, OJSC Dniprokoks, OJSC Dniproshyna, OJSC Dniprotyazhmash, OJSC Dnipro Paint and Varnish Plant, CJSC Dnipro Oil Extraction Plant. The substances that pollute the air most in the city are nitrogen dioxide, formaldehyde, benz(a)pyrene. The average annual concentration of formaldehyde is about 3 MPC, benz(a)pyrene—2 MPC.

Odesa. About 1 million people live in the city. The sectors with the greatest influence on the quality of atmospheric air are the chemical and oil refining industries, mechanical engineering, electric power, and building materials. The total number of enterprises that in the process of activity affect the state of atmospheric air in the Odesa region is about 3 thousand units. During 2018, 18.3 thousand tons of harmful substances from stationary sources entered to the air basin of the region, which is 50% more than in 2017.

The enterprises, that emit the largest amount of pollutants, are CJSC Odescement, TPP-1, Odesa Commercial Sea Port, Odesa Refinery. The most polluted part of the city is the northern one, where the most powerful industrial enterprises are concentrated. The air in the city is the most polluted by formaldehyde (5 MPC), nitrogen dioxide (2 MPC), phenol (2 MPC), hydrogen fluoride (2 MPC).

Kharkiv. The population is about 1.5 mln people. Engineering is the most influential industry on air pollution. The main source of air pollution in the city is road transport. Among stationary sources of air pollution are energy enterprises (TPP-3, TPP-5), engineering (KhTZ, VA Malyshev Plant), and the construction industry. According to the Main Statistics Directorate on the Kharkiv region, emissions of pollutants into the air from stationary sources in 2018 amounted to 44.7 thousand tons (in 2017—45 thousand tons, in 2016—100.2 thousand tons). The decrease in emissions of pollutants into the air is associated with a decrease in the production volumes of enterprises in the energy sector, in particular, the Zmievskaya TPP of PJSC Centrenergo (in 2018 10.5 thousand tons, in 2017—34.1 thousand tons).

5   Air Pollution Monitoring System in Ukraine

Monitoring of atmospheric air is a component of the state system of environmental monitoring; it is carried out with the aim of obtaining, collecting, processing, storing and analyzing information on the state of atmospheric air and developing scientifically based recommendations for making decisions in the field of atmospheric air protection.

The subjects of atmospheric air monitoring are the Ministry of Energy and Environmental Protection of Ukraine, the Ministry of Health Protection, the State Service for Surveillance of Situations, the State Agency for the Management of the Exclusion Zone, the executive agency of the Autonomous Republic of Crimea on environmental protection, regional, Kyiv city state administration, executive agencies of city councils.

According to [8], all the above agencies, except the Ministry of Energy and Environmental Protection of Ukraine, have the right to establish observation points and monitor the levels of pollutants. In practice, the above monitoring entities use the data of the State Hydrometeorological Service and the reporting data provided by industrial enterprises.

The subjects of atmospheric air monitoring place observation posts, conduct monitoring of the concentrations of pollutants listed below:

- Sulfur dioxide
- Nitrogen dioxide and nitrogen oxides
- Benzene
- Carbon monoxide
- Plumbum
- Particulate matter PM10

- Particulate matter PM2.5
- Arsen
- Cadmium
- Mercury
- Nickel
- Benz(a)pyrene
- Ozone

Enterprises, institutions and organizations whose activities lead or may lead to a deterioration in the state of atmospheric air can establish observation posts and monitor the concentrations of pollutants listed below:

- Ammonia
- Aniline
- Hydrogen chloride
- Hydrogen cyanide
- Iron and its compounds (in terms of iron)
- Nitric acid
- Sulfuric acid
- Xylene
- Volatile organic compounds
- Manganese and its compounds (in terms of manganese dioxide)
- Copper and its compounds (in terms of copper)
- Soot
- Hydrogen sulfide
- Carbon disulphide
- Phenol
- Hydrogen fluoride
- Chlorine
- Chloralines
- Chromium and its compounds (in terms of chromium)
- Zinc and its compounds (in terms of zinc)

The number of observation posts and their placement for assessment is determined in the state monitoring program in the field of atmospheric air protection for each zone and metropolitan area, in accordance with the procedure established by the Ministry of Internal Affairs in coordination with the Ministry of Energy and Environmental Protection of Ukraine.

To ensure the accuracy of measuring devices, all subjects of atmospheric air monitoring control the concentration of pollutants, carry out atmospheric air quality assessments, provide calibration and maintenance of measuring instruments used to monitor atmospheric air.

The main method for determining pollutant concentrations is air sampling at stationary observation posts. The number of posts is determined by the size of the city and the characteristics of the industrial structure. It can range from one post for cities with a population of less than 50,000 inhabitants, to 20 posts for cities

Table 4 The principle of building an environmental monitoring network in urban metropolitan areas

Population	Number of posts by RD 52.04.186–89	Number of posts by Directive 2008/50/EU
<50 k	1	1
50–100 k	2	1
100–200 k	2–3	1
200–500 k	3–5	2 (250–499 k)
500 k–1 mln	5–10	2 (500–749 k)
>1 mln	10–20	2 (750–999 k) 3 (1–1,9 mln) 4 (2–3,749 mln) 6 (3,750–5,999 mln) 7 (>6 mln)

with a population of over one million. In 2016, the country had 129 posts in 39 cities. Most of all, 16 posts—in Kyiv, 10 posts—in Kharkiv, 8—in Odesa, 6—in Dnipro. Large industrial centers—Zaporizhia, Kryvyi Rih, Mariupol—each had 5 observation posts, while for the majority of regional centers their number did not exceed 4.

A comparative analysis of the construction of a network for monitoring atmospheric air in accordance with the requirements of RD 52.04.186–89 and Directive 2008/50/EC is given in Table 4. The required number of posts in Ukraine in accordance with the requirements of RD 52.04.186–89 and Directive 2008/50/EU are given in Table 5.

Sampling is carried out at certain time intervals (periods) in accordance with one of the four observation programs: complete, incomplete, shortened or daily. The full program provides four measurements during the day: at 01:00, 7:00, 13:00, 19:00 local time; incomplete—three: at 07:00, 13:00, 19:00; abbreviated—two: at 07:00, 13:00; the daily program provides for continuous monitoring.

Table 5 The number of environmental monitoring posts in urban agglomerations of Ukraine

Population	Number of cities with corresponding population	Total number of posts by RD 52.04.186–89	Total number of posts by directive 2008/50/EU
<50 k	373	373	373
50–100 k	44	88	44
100–200 k	13	26–39	13
200–500 k	24	72–120	48
500 k–1 mln	5	25–50	10
>1mln	3	30–60	7
Total number of posts		**614–730**	**495**

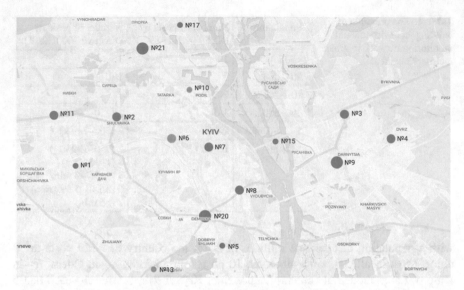

Fig. 9 Atmospheric air quality monitoring network of the Central Geophysical Observatory (a division of the Hydrometeorological Center) in Kyiv

Monitoring of concentrations of dust, sulfur dioxide, carbon monoxide, nitrogen dioxide, lead and its inorganic compounds, benz(a)pyrene, formaldehyde and radioactive substances are mandatory. Other substances may be included in the observation program by decision of local authorities in accordance with the specifics of the environmental situation.

Now in Ukraine there is no modern information-measuring air monitoring system. Data of the concentration of pollutants publishes Sreznevsky Central Geophysical Observatory while the data are published with a delay of 24 h. At present, the observatory publishes data from posts No. 3, 5, 7, 20 (Fig. 9). The observatory website (http://cgo-sreznevskyi.kiev.ua/) has data for the last 5 days.

In Fig. 9, blue color indicates the posts at which monitoring is carried out 4 times a day, yellow—2. Concentrations of suspended solids, sulfur dioxide, carbon monoxide and nitrogen dioxide are measured in Kyiv at more than ten posts simultaneously. But the concentration of soluble sulfates, nitric oxide and hydrogen sulfide are determined at 3 or less posts. The larger the circle, the greater the number of substances that are monitored.

For example, Fig. 10 shows the average daily concentration of formaldehyde in Kyiv (post № 7 of the Sreznevsky Central Geophysical Observatory). Data for analysis was collected by copying from the site of the observatory in the period from October 27, 2017 to January 25, 2020. 1988 measurements were obtained for this period. Since sampling was carried out 4 times a day, 497 daily average concentrations of formaldehyde were obtained. Table 6 shows the distribution of average daily concentrations of formaldehyde to normative.

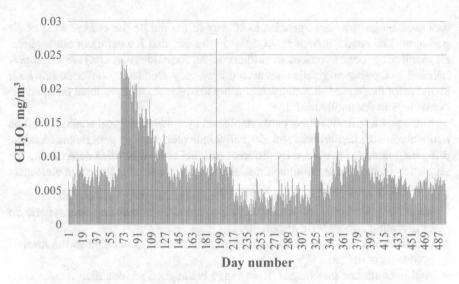

Fig. 10 Average daily concentration of formaldehyde in Kyiv in the period from October 27, 2017 to January 25, 2020

Table 6 Volume of average daily concentrations of formaldehyde according to normative

Pollutant	Safety	Slightly dangerous	Moderately dangerous	Dangerous	Very dangerous
Formaldehyde	24	148	278	45	2

In further studies, we will analyze in detail air pollution according to the collected data.

6 World Air Monitoring Approaches

The main structural elements of international air quality monitoring are:

- availability of an extensive system of monitoring stations;
- availability of a methodology for measuring key indicators of air quality along with monitoring of the meteorological condition;
- system for collecting, analyzing and transmitting data on the state of atmospheric air quality;
- availability of a strategy to support and develop a monitoring system;
- availability of communication tools for the quality of atmospheric air.

For assessing the quality of atmospheric air abroad, in particular in Europe, the air quality index (AQI) is used [1, 2, 28, 29]. By its nature, the AQI is a communication tool used by government agencies to convey to the public the current state of air pollution. The need for such an AQI lies in the fact that indicators of atmospheric air monitoring (concentrations of sulfur dioxide, formaldehyde, etc.) are incomprehensible to the general public and, accordingly, they need to be converted into such an indicator that would show the relationship between the observational data and the consequences for public health.

Since possible health effects are established by epidemiological studies based on national research institutions, and air quality indicators vary by geographic location, different countries are guided by different national standards when determining the air quality index. But at the same time, all indices have similar structural elements:

- calculation of the AQI is carried out on the basis of average values of the concentration of pollutants for a single period, obtained by monitoring atmospheric air or modeling atmospheric dispersion;
- concentration and time of fixation of this concentration is taken as the level of pollutants in the air;
- AQI is combined into ranges. Each range is assigned an identifier, a color code and recommendations for the public to protect their own health;
- AQI is built in the manner in which it is assumed that an increase in the index will indicate that a significant part of the population will face serious health consequences.

According to the EPA standard, the following substances can be used to determine the AQI: SO_2, NO_2, CO, O_3, PM2.5, PM10. Table 7 shows the ranges of concentrations of pollutants and the corresponding ranges of AQI. The formula for determining AQI is as follows:

$$I = \frac{I_{max} - I_{min}}{C_{max} - C_{min}} \cdot (C - C_{min}) + I_{min}, \tag{2}$$

Table 7 Ranges of AQI values and air pollutant concentrations

AQI	NO_2 (ppb)	SO_2 (ppb)	CO (ppm)	PM_{10} ($\mu g/m^3$)	$PM_{2.5}$ ($\mu g/m^3$)	O_3 (ppb)
I_{min}–I_{max}	C_{min}–C_{max}	C_{min}–C_{max}	C_{min}–C_{max}	C_{min}–C_{max}	C_{min}–C_{max}	C_{min}–C_{max}
0–50	0–53[*]	0–35[*]	0–4.4[**]	0–54[***]	0–12[***]	0–54[**]
51–100	54–100[*]	36–75[*]	4.5–9.4[**]	55–154[***]	12.1–35.4[***]	55–70[**]
101–150	101–360[*]	76–185[*]	9.5–12.4[**]	155–254[***]	35.5–55.4[***]	71–85[**] 125–164[*]
151–200	361–649[*]	186–304[*]	12.5–15.4[**]	255–354[***]	55.5–150.4[***]	86–105[**] 165–204[*]
201–300	650–1249[*]	305–604[***]	15.5–30.4[**]	355–424[***]	105.5–250,4[***]	106–200[**] 205–404[*]
301–400	1250–1649[*]	605–804[***]	30.5–40.4[**]	425–504[***]	250.5–350.4[***]	405–504[*]
401–500	1650–1049[*]	805–1004[***]	40.5–50.4[**]	505–604[***]	350.5–500.4[***]	505–604[*]

Note *—the average concentration of the pollutant over 1 h; **—the average concentration of the pollutant for 8 h; ***—the average value of the concentration of the pollutant for 24 h

where I is the current value of AQI; I_{max} is the maximum value of AQI for the current concentration range of the pollutant; I_{min} is the minimum AQI value for the current range of the pollutant's concentration; C_{max} is the maximum value of the current concentration range of the pollutant; C_{min} is the minimum value of the current concentration range of the pollutant; C is the current concentration of the pollutant.

Table 8 shows the hazard level data characterizing each of the corresponding AQI ranges.

Figure 11 shows a screenshot of waqi.info, which shows air pollution by various pollutants, as well as AQI.

European institutions use AQI as a research and communication tool. At the same time, the European agencies also use the Common Air Quality Index (CAQI), which allows you to display air quality in European cities and is divided into 3 different indices, which differ in time intervals:

- The hourly index describes air quality based on hourly values and is updated every hour;
- The daily index is responsible for the general air quality of the previous day, is based on daily values and is updated once a day;
- The annual index shows AQI throughout the year and is compared with European air quality standards. This indicator is based on an average level for a year, in accordance with annual limit values, and is updated once a year.

On November 16, 2017, the European Environment Agency (EEA) launched the European Air Quality Index (Fig. 12).

This AQI allows real-time monitoring of air quality indicators in those countries that have implemented real-time data transfer protocols.

Table 8 Caution regarding AQI levels

AQI	Designation	Warning
0-50	good	Air quality is satisfactory, air pollution is negligible (in normal limits)
51-100	satisfactorily	Air quality is acceptable, but some pollutants can be dangerous for people who are especially sensitive to polluted air
101-150	bad for sensitive groups	An effect on a particularly sensitive group of individuals may be observed. No visible effect on the average resident
151-200	bad	Everyone can feel the consequences for their health. A particularly sensitive group may feel more serious consequences
201-300	very bad	Health hazard from emergency. Probably there will be an effect on the entire population
300+	dangerous	Health hazard, everyone can feel serious consequences for their health

Fig. 11 Screenshot of waqi.info

Fig. 12 Visualization of European Air Quality Index (February 24, 2020)

The index uses more than 2,000 air quality control stations in Europe. All of them belong to the Copernicus atmosphere monitoring network.

The index estimates air quality for 4 indicators: particulate matter (PM2.5 and PM10), ground-level ozone (O_3), nitrogen dioxide (NO_2) and sulfur dioxide (SO_2). Each of these indicators is evaluated in accordance with the standards approved by the European Union Directives. Since the standards distinguish between indicators in the long term (annual cycle) and in the short term (hours and days), the index provides information on air quality only in the short term.

AQI updates data every 6 h, but has the ability to display data in any chronological interval between 0 to 48 h. There are also cases when data from analyzers is not received on time. In order to solve both problems, the European Agency uses approximation methods to model data for such cases.

The method itself differs depending on the measured indicator.

The differential method is used for nitrogen dioxide NO_2 and particulate matter PM2.5 and PM10 (the value is obtained by simulating the Copernicus system with the addition or subtraction of the correction difference. The latter is averaging the difference between previous measurements and system-modeled values obtained to the same hour for at least 3 of the 4 previous days).

A multiplicative method is used for ground-based ozone O_3 (the value is obtained by modeling the Copernicus system with the addition of a correction factor. This factor is the average ratio between the previously measured values and the models obtained at the same hour for at least 3 of 4 previous days).

For sulfur dioxide, these methods are not applicable.

The system was developed jointly by the European Environment Agency and the European Commission's Directorate for the Environment. The cost of the project was not disclosed. From a technical point of view, the map is an adaptation of the JavaScript library of the Mapbox service. Other libraries were also used for additional functions. The index does not allow downloading data for analysis and serves only as a communication tool. But it contains links to the primary data, which can be downloaded on another resource.

7 Disadvantages of Air Quality Monitoring System in Ukraine

The organization and methodology of monitoring air quality in Ukraine does not comply with EU standards.

The requirements for the number of observation posts in the settlements of Ukraine exceed similar EU standards. If in Ukraine for a city with a population of about 3 million people it is necessary to have from 10 to 20 posts, then the EU directive for a city of the same size establishes their minimum number of 4 posts. In Ukraine, observing programs can differ significantly in individual posts, but in the EU, the principle of measuring all substances at all posts is valid. Moreover, even the above requirements of RD 52.04.186–89 in terms of the number of posts in places are not fulfilled [30].

The list of substances observed in Ukraine does not meet current needs. Direct prescribing of this list in regulatory documents led to a paradoxical situation. In Kyiv, for example, about 20 pollutants are monitored. At the same time, there is no separation of suspended particles into PM10 and PM2.5; a number of aggressive and common substances remain without attention: ozone, benzene, arsenic, and mercury. In the EU, an observing program is formed on the basis of threshold levels, the excess of which determines the need to introduce a certain type of monitoring. This eliminates the need to measure a large amount of substances, focusing instead on key pollutants. Methods for measuring concentrations that are used today in Ukraine also need to be improved [31, 32].

The use of indicative measurement or modeling is poorly regulated by the regulatory framework of Ukraine. As a result, the existing monitoring system has limited data on the state of air pollution throughout the territory, as well as on the long-term dynamics of indicators. For example, in Kyiv, emissions from motor vehicles cause a significant level of air pollution. Depending on the characteristics of the motorway, traffic flows, weather, land use, development and vegetation, their distribution is different. Even within a few hundred meters, significant differences in concentrations of harmful substances can be observed. To account for all of these factors, in addition to fixed observations, it is also necessary to simulate pollution.

Another drawback of the existing monitoring system is the measurement at specified time intervals(01:00, 07:00, 13:00, 19:00). This makes it impossible to fix the maximum values if they are observed in a different period of time, and also affects the accuracy of averaging. For comparison, the EU practices constant and, in some cases, random observations.

Hygienic standards of air quality, which are used in Ukraine, are also imperfect. In recent studies, there are more and more mixed assessments of the use of MPC. They are criticized for not taking into account both direct and indirect effects, as well as the use of organoleptic or reflex signs that cannot indicate a health disorder. The standards for some substances are developed in an accelerated manner and generally do not carry information about the possible consequences of exposure. Given this, the reliability of the MPC is doubtful. A better indicator used in international practice is risk assessment. Unlike MPC, it: (a) takes into account only the direct effect of substances and the sensitivity of various population groups; (b) provides that some substances have a thresholdless carcinogenic or mutagenic effect; and (c) takes into account the possibility of simultaneous harmful effects of several substances.

It is also important to introduce environmental safety standards to prevent the negative impact of pollution not only on human health, but also on the environment. Despite the fact that the procedure for developing these standards has been defined by law for more than 15 years, the relevant standards have not been approved by the Ministry of Energy and Environmental Protection of Ukraine.

Finally, in Ukraine there is no regulatory framework to inform the public about the quality of atmospheric air. Monitoring data does not apply to open data. Access to them is regulated within the framework of the general legislation on access to public information. In Kyiv, for example, some information on the concentration of pollutants is published by the Sreznevsky Central Geophysical Observatory. However, they are not complete and are stored on the site only for a few days. In many other cities, even such information is absent.

A way out of this essence could be stations of public monitoring of air quality. In Ukraine, a network of Eco-city air quality posts is being widely used (https://eco-city.org.ua) (Fig. 10). The technical features of the stations of the Eco-city network, the existing network and its advantages will be discussed in the following works.

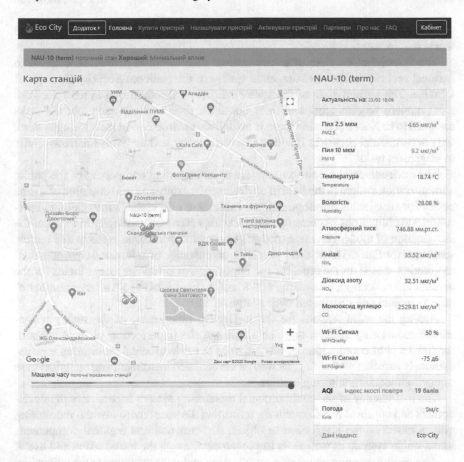

Fig. 13 Eco-city user account

However, now public monitoring stations can measure the concentrations: PM1, PM2.5, PM10, CO, CO_2, NH_3, NO_x, formaldehyde, VOC and others. Such network of stations can become the main system of indicative monitoring of air quality in Ukraine (Fig. 13).

8 Conclusions

The development of an effective environmental monitoring system is in line with European and global approaches to environmental management, including the requirements and guidelines of the Association Agreement between Ukraine and the European Union. The creation of such a system will significantly expand the possibilities of international cooperation of Ukraine in the field of environmental protection and contribute to bringing the state of the environment in line with European and world requirements.

The unsatisfactory state of environmental monitoring is defined as a threat to national security in the environmental sphere in the National Security Strategy of Ukraine. The current state of the state environmental monitoring system does not meet modern requirements and does not allow to effectively carry out the tasks assigned to it. Now in Ukraine there is no holistic environmental monitoring system, only departmental networks function in a certain way, and they solve narrow-profile management tasks.

The environmental monitoring system as an important component of the public administration system in the field of environmental safety and the formation of a state policy for sustainable development, the fulfillment of Ukraine's international environmental obligations requires fundamental improvement, especially in terms of the introduction of modern technologies of geographic information systems and remote sensing of the Earth. The primary task is the organization of automated environmental monitoring, based on the creation and application of computer technologies for the operational collection, processing and transmission of data from a large number of objects remote and distributed over a large territory.

Among the main reasons for the ineffective functioning of the state environmental monitoring system, one can single out the imperfection of legal support, the low level of coordination of the activities of environmental monitoring entities, extremely insufficient funding, as well as the outdated instrument and technical base of environmental monitoring entities.

The inefficiency of the environmental monitoring system necessitates the development and approval of a Concept for reforming the state environmental monitoring system based on the integration of object, departmental and regional components into a single system, as well as its improvement through the introduction and use of modern geoinformation and communication technologies to automate the collection, processing and analysis of the results of observations.

References

1. Jamil, M.S., Jamil, M.A., Mazhar, A., Ikram, A., Ahmed, A., Munawar, U.: Smart environment monitoring system by employing wireless sensor networks on vehicles for pollution free smart cities. Proc. Eng. **107**, 480–484 (2015). https://doi.org/10.1016/j.proeng.2015.06.106
2. Maslyiak, Y., Pukas, A., Voytyuk, I., Shynkaryk, M.: Environmental monitoring system for control of air pollution by motor vehicles. In: 2018 XIV-th International Conference on Perspective Technologies and Methods in MEMS Design (MEMSTECH), 18–22 April 2018, pp. 250–254. Lviv, Ukraine. https://doi.org/10.1109/memstech.2018.8365744
3. Shaban, K.B., Kadri, A., Rezk, E.: Urban air pollution monitoring system with forecasting models. IEEE Sens. J. **16**(8), 2598–2606 (2016). https://doi.org/10.1109/JSEN.2016.2514378
4. Kol'tsov, M., Shevchenko, L.: Monitoring air quality: Ukrainian and international experience [Analytical note]. Kyiv, OO "Open Society Fund", p. 13 (2018)
5. Directive 2008/50/EC of the European Parliament and the Council of 21 May 2008. On Ambient air quality and cleaner air for Europe. Online: https://www.ecolex.org/details/legisl

ation/directive-200850ec-of-the-european-parliament-and-of-the-council-on-ambient-air-qua
lity-and-cleaner-air-for-europe-lex-faoc080016/

6. Putrenko, V.V., Panshynska, N.M.: The use of remote sensing data for modeling air quality in the cities. In: ISPPS Annals of the Photogrammetry, Remote Sensing and Spatial Information Sciences, vol. IV-5/W1, pp. 57–62 (2017). https://doi.org/10.5194/isprs-annals-iv-5-w1-57-2017

7. Babiy, A.P., Kharytonov, M.M., Gritsan, N.P.: Connection between emissions and concentrations of atmospheric pollutants. In: Melas, D., Syrakov, D. (eds.) Air Pollution Processes in Regional Scale. NATO Science Series (Series IV: Earth and Environmental Sciences), vol. 30, pp. 11–19. Springer, Dordrecht (2003). https://doi.org/10.1007/978-94-007-1071-9_2

8. Resolution of the Cabinet of Ministers of Ukraine of 08/14/2019. Some issues of state monitoring in the field of atmospheric air protection. Online: https://zakon.rada.gov.ua/laws/show/827-2019-%D0%BF

9. Kuchansky, A., Biloshchytskyi, A., Andrashko, Y., Vatskel, V., Biloshchytska, S., Danchenko, O., Vatskel, I.: Combined models for forecasting the air pollution level in infocommunication systems for the environment state monitoring. In: 2018 IEEE 4th International Symposium on Wireless Systems within the International Conferences on Intelligent Data Acquisition and Advanced Computing Systems (IDAACS-SWS), 20–21 Sept. 2018, pp. 125–130. Lviv, Ukraine (2018). https://doi.org/10.1109/idaacs-sws.2018.8525608

10. Shparyk, Y.S., Parpan, V.I.: Heavy metal pollution and forest health in the Ukrainian Carpathians. Env. Poll. **130**(1), 55–63 (2004). https://doi.org/10.1016/j.envpol.2003.10.030

11. Birmili, W., Schepanski, K., Ansmann, A., Spindler, G., Tegen, I., Wehner, B., Nowak, A., Reimer, E., Mattis, I., Muller, K., Bruggemann, E., Gnauk, T., Hermann, H., Wiedensohler, A., Althausen, D., Schladitz, A., Tuch, T., Loschau, G.: An case of extreme particulate matter concentrations over Central Europe caused by dust emitted over the southern Ukraine. Atm. Chem. Phy. **8**, 997–1016 (2008). https://doi.org/10.5194/acp-8-997-2008

12. Shupranova, L.V., Khlopova, V.M., Kharytonov, M.M.: Air pollution assessment in the dnepropetrovsk industrial megapolice of Ukraine. In: Steyn, D., Builtjes, P., Timmermans, R. (eds.) Air Pollution Modeling and its Application XXII. NATO Science for Peace and Security Series C: Environmental Security, pp. 101-104. Springer, Dordrecht (2014). https://doi.org/10.1007/978-94-007-5577-2_17

13. Palekhov, D., Schmidt, M., Pivnyak, G. (2008) Standards and thresholds for EA in highly polluted areas—the approach of Ukraine. In: Schmidt, M., Glasson, J., Emmelin, L., Helbron, H. (eds.) Standards and Thresholds for Impact Assessment. Environmental Protection in the European Union, vol 3, pp. 33–48 (2008). https://doi.org/10.1007/978-3-540-31141-6_3

14. Zvyagintsev, A.M., Blum, O.B., Glazkova, A.A., Kotel'nikov, S.N., Kuznetsova, I.N., Lapchenko, V.A., Lezina, E.A., Miller, E.A., Milyaev, V.A., Popikov, A.P., Semutnikova, E.G., Tarasova, O.A., Shalygina, I.Yu.: Air pollution over European Russia and Ukraine under the hot summer conditions of 2010. Izvestiya, Atm. Ocean. Phy. **47**, 699–707 (2011). http://doi.org/10.1134/S0001433811060168

15. Bakharev, V., Marenych, A. (2017). The key aspects of atmospheric air ecological monitoring concept formation at the urban systems level. Env. Prob. **1**(2), 25–25. http://ena.lp.edu.ua:8080/handle/ntb/39434

16. Makarenko, N., Budak, O.: Waste management in Ukraine: municipal solid waste landfills and their impact on rural areas. Annals Agr. Sci. **15**(1), 80–87 (2017). https://doi.org/10.1016/j.aasci.2017.02.009

17. Plakhotnik, V.N., Onyshchenko, JuV, Yaryshkina, L.A.: The environmental impacts of railway transportation in the Ukraine. Trans. Res. Part D: Trans. Env. **10**(3), 263–268 (2005). https://doi.org/10.1016/j.trd.2005.02.001

18. Brody, M., Caldwell, J., Golub, A.: Developing risk-based priorities for reducing air pollution in urban settings in Ukraine. J. Toxicol. Env. Health **70**(3–4), 352–358 (2007). https://doi.org/10.1080/15287390600885021

19. Sindosi, O.A., Katsoulis, B.D., Bartzokas, A.: An objective definition of air mass types affecting Athens, Greece; The corresponding atmospheric pressure patterns and air pollution levels. Env. Technol. **24**(8), 947–962 (2003). https://doi.org/10.1080/09593330309385633

20. Mokin, V.B. (2007). Development of the geoinformation system of the state ecological moni-
    toring. In: Morris, A., Kokhan, S. (eds.) Geographic Uncertainty in Environmental Security.
    NATO Science for Peace and Security Series C: Environmental Security, pp. 153–165. Springer,
    Dordrecht (2007). https://doi.org/10.1007/978-1-4020-6438-8_9
21. Popov, O., Iatsyshyn, A., Kovach, V., Artemchuk, V., Taraduda, D., Sobyna, V., Sokolov, D.,
    Dement, M., Hurkovskyi, V., Nikolaiev, K., Yatsyshyn, T., Dimitriieva, D.: Physical features
    of pollutants spread in the air during the emergency at NPPs. Nuc. Radia. Saf. **84**(4), 88–98
    (2019). https://doi.org/10.32918/nrs.2019.4(84).11
22. Popov, O., Iatsyshyn, A., Kovach, V., Artemchuk, V., Taraduda, D., Sobyna, V., Sokolov, D.,
    Dement, M., Yatsyshyn, T., Matvieieva, I.: Analysis of possible causes of NPP emergencies to
    minimize risk of their occurrence. Nucl. Radia. Saf. **81**(1), 75–80 (2019). https://doi.org/10.
    32918/nrs.2019.1(81).13
23. Popov, O.O., Iatsyshyn, A.V., Kovach, V.O., Artemchuk, V.O., Kameneva, I.P., Taraduda, D.V.,
    Sobyna, V.O., Sokolov, D.L., Dement, M.O., Yatsyshyn, T.M.: Risk assessment for the popu-
    lation of Kyiv, Ukraine as a result of atmospheric air pollution. J. Health Poll. **10**(25), 200303
    (2020). https://doi.org/10.5696/2156-9614-10.25.200303
24. Zaporozhets, A.: Analysis of control system of fuel combustion in boilers with oxygen Sensor.
    Peri. Polytec. Mech. Eng. **63**(4), 241–248 (2019). https://doi.org/10.3311/PPme.12572
25. Zaporozhets, A. (2019). Development of software for fuel combustion control system based
    on frequency regulator. In: CEUR Workshop Proceedings, vol. 2387, pp. 223–230. http://ceur-
    ws.org/Vol-2387/20190223.pdf
26. Babak, V.P., Mokiychuk, V.M., Zaporozhets, A.A., Redko, A.A. (2016). Improving the effi-
    ciency of fuel combustion with regard to the uncertainty of measuring oxygen concentration.
    East. Eur. J. Ent. Technol. **6**(8(84)), 54–59 (2016). https://doi.org/10.15587/1729-4061.2016.
    85408
27. Babak, S., Babak, V., Zaporozhets, A., Sverdlova, S. (2019). Method of statistical spline func-
    tions for solving problems of data approximation and prediction of objects state. In: CEUR
    Workshop Proceedings, vol. 2353, pp. 810–821. http://ceur-ws.org/Vol-2353/paper64.pdf
28. Xiaojun, C., Xianpeng, L., Peng, X.: IOT-based air pollution monitoring and forecasting system.
    In: 2015 International Conference on Computer and Computational Sciences (ICCCS), 27–29
    Jan, 2015, pp. 257-260. Noida, India (2015). https://doi.org/10.1109/iccacs.2015.7361361
29. Plakhotnij, S.A., Klyuchko, O.M., Krotinova, M.V.: Information support for automatic indus-
    trial environment monitoring systems. Elec. Contr. Sys. **1**(47), 29–34 (2016). https://doi.org/
    10.18372/1990-5548.47.10266
30. Kharchenko, V., Prusov, D.: Analysis of unmanned aircraft systems application in the civil
    field. Transport **27**(3), 335–343 (2012). https://doi.org/10.3846/16484142.2012.721395
31. Zaporozhets, A., Eremenko, V., Serhiienko, R., Ivanov, S. (2019). Methods and Hardware for
    Diagnosing Thermal Power Equipment Based on Smart Grid Technology. In: Shakhovska,
    N., Medykovskyy, M. (eds.) Advances in Intelligent Systems and Computing III. CSIT 2018.
    Advances in Intelligent Systems and Computing, vol. 871, pp. 476–489. Springer, Cham (2019).
    https://doi.org/10.1007/978-3-030-01069-0_34
32. Zaporozhets, A.O., Redko, O.O., Babak, V.P., Eremenko, V.S., Mokiychuk, V.M.: Method
    of indirect measurement of oxygen concentration in the air. Naukovyi Visnyk Natsionalnoho
    Hirnychoho Universytetu **5**, 105–114 (2018). https://doi.org/10.29202/nvngu/2018-5/14

Modeling of the Process of Optimization of Decision-Making at Control of Parameters of Energy and Technical Systems on the Example of Remote Earth's Sensing Tools

Oleksandr Maevsky⬤, Volodymyr Artemchuk⬤, Yuri Brodsky⬤,
Igor Pilkevych⬤, and Pavlo Topolnitsky⬤

Abstract The authors study the process of decision-making optimization in the control of the spacecraft onboard systems. To ensure the continuous operation of a remotely controlled complex technical system, it is necessary, on the basis of an analysis of the state of the onboard systems, to formulate control effects, the absence of which could lead to the system's failure to fulfill its tasks or system failure. In order to prevent such situations, an approach based on a simulation model is proposed, the use of which will reduce the risk of accidents in the onboard systems of the spacecraft. The proposed model is represented by factor space. The state of the onboard parameters of the spacecraft at different points in time is matched by the set of points that form the decision-making surface in this factor space. The basic stages of forming the optimal trajectory on the decision surface, which are approximated by numerical methods, are given and described. Using the actual values of the parameters obtained in a 15-minute data communication session from the board of the artificial satellite Earth "Ocean—1", a decision-making surface was constructed. The equation of the optimal trajectory on the created surface is obtained. The simulation results will be used to develop emergency management and control systems.

Keywords Remote earth's sensing (RES) · Spacecraft · Onboard parameters ·
Optimal trajectory · Decision surface · Factor space · Simulation model

O. Maevsky · Y. Brodsky · P. Topolnitsky
Zhytomyr National Agroecological University, Zhytomyr, Ukraine

V. Artemchuk (✉)
Pukhov Institute for Modelling in Energy Engineering of NAS of Ukraine, Kyiv, Ukraine
e-mail: ak24avo@gmail.com

I. Pilkevych
Zhytomyr Military Institute, Zhytomyr, Ukraine

V. Babak et al. (eds.), *Systems, Decision and Control in Energy I*, Studies in Systems,
Decision and Control 298, https://doi.org/10.1007/978-3-030-48583-2_7

1   Introduction

In the course of its functioning, society influences the environment through wars and local conflicts, the misuse of natural resources, and accidents at industrial sites. In addition, natural disasters such as earthquakes, tsunamis, hurricanes, and more. also lead to unfavorable living conditions of mankind [1].

All this leads to the need to use environmental monitoring tools [2, 3]. For this purpose, various technical means, equipped with the equipment of registration and analysis of environmental factors are used. Today we have a tendency to increase the autonomy of these tools to improve their efficiency. In this case, there is a need for comprehensive control of the operation of these facilities to ensure their quality functioning on the one hand, and to eliminate the negative impact of these tools on the environment in the event of failure of the technical monitoring equipment. Because of their inherent benefits, one of the most commonly used environmental monitoring tools is the remote Earth's sensing (Fig. 1).

2   Literature Analysis and Problem Statement

In general, the task of constructing mathematical models of processes of different physical nature according to the obtained experimental data is to determine the parameters of the approximating function.

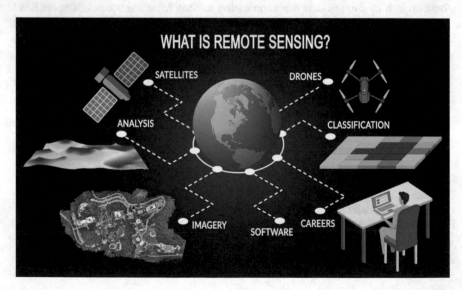

Fig. 1   Remote Earth's sensing tools [4]

However, the rather complex nature of the dynamics of changing the parameters of the onboard systems of the means of the RES leads to complex nonlinear dependencies both in the form of approximating functions and in the form of differential equations and their systems. At the same time, the construction of an adequate mathematical model based on the experimental data provides the necessary reproduction of the dynamics of the studied process with a minimum total error and, as a consequence, increases the efficiency of control of the systems of means of RES.

When using mathematical models represented by systems of differential equations, there is a complication of computational algorithms.

For polynomial mathematical models, there is an increase in simulation errors by increasing the number of coefficients (in the corresponding functional basis), which in turn have an error in their calculation.

Methods of construction and peculiarities of creation of the mentioned mathematical models reproduced in the works of a number of scientists: S. Kuzmin, V. Baranov, E. Lviv, Yu. Linnik, V. Mudrov, V. Kushka, V. Kuntsevich, O. Kukusha, R. Kalman, E. Sage, J. Melsa, V. Medich, I. Shapiro, D. Kahaner [5–9] and others.

To prevent the occurrence of risky situations, these polynomial mathematical models and mathematical models presented by systems of differential equations, as well as stochastic mathematical models are used [10–13].

In addition, it is worth noting a number of works on optimization of decision-making on the management of complex technical systems [14–21].

The approach proposed in the article to create a mathematical model for preventing the occurrence of risky situations, unlike these types of models, does not require extrapolation, but is focused on the creation of the factor space of the parameters of the onboard systems of the RES means within the specified limits and the optimal control of the critical values of the control parameters of the RES systems.

3 Purpose and Objectives of the Study

To ensure the continuous operation of a remotely controlled complex technical system, it is necessary to formulate, on the basis of an analysis of the state of the onboard systems, control effects, the absence of which may lead to the system's failure to fulfill its tasks or system failure. In order to prevent such situations, an approach based on a simulation model is proposed, the use of which will reduce the risk of accidents in the onboard systems of the RES.

During the operation of spacecraft (SC) in the orbit of the onboard and ground control systems, the conduct and performance of a whole series of onboard mechanisms (current and voltage on the rising solar cells, heavy power plants, etc.) in the mandatory range for continuous operation of the spacecraft and exchange of information from the flight control center. These questions are used by the telemetry control system, which is created in the control of the spacecraft. The onboard SC system is used and specified throughout the work.

However, due to various factors disclosed by the physical nature, the parameters of the onboard systems of the RES can reach critical or unacceptable values, which can be negatively observed in the mode of operation of the spacecraft, provoke the occurrence of an emergency situation and loss of spacecraft. In this case, the onboard systems accept a uniform spacecraft, always needing to optimize the operation of the onboard system and return the parameters to the range of possible meaning.

Thus, an important and relevant scientific and practical task is the efficiency of control of the onboard systems of the spacecraft. In this regard, we propose a model that optimizes the process of restoring the onboard parameters of the spacecraft within the operating range.

The proposed simulation model is a factor space. The states of the onboard parameters of the spacecraft at different times are matched by the set of points that form the decision-making surface in this factor space.

If critical or unacceptable values of the onboard parameters of the spacecraft are reached, the decision-making system of the spacecraft returns them to nominal limits, which corresponds to the movement of the end of the radius vector from point to point on the surface of decision making in factor space (Fig. 2).

To improve decision-making efficiency, it is necessary that the radius vector moves from point to point along optimal trajectories belonging to the decision surface. In this case, the optimal trajectories will be "straight" on the decision surface that has Riemann geometry. In this case, it is necessary to establish the geometry of the decision surface and find the equation of optimal trajectories on it, which is the main purpose of the conducted research.

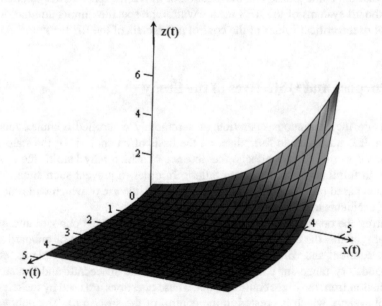

Fig. 2 Decision-making surface in factor space (units are given conventionally)

4 Research Methods

This study begins with a brief analysis of publications on modeling the decision-making optimization process when managing the parameters of technical systems, and focuses mainly on developing a simulation model.

The following methods were used in this study: the method of comparative analysis; mathematical modeling; simulation modeling; numerical methods.

5 Research Results

Modeling.

As is known, the equation of optimal trajectories is obtained after equating to zero all curvilinear components of the acceleration of a point moving on a given surface.

$h_k = \sqrt{g_{kk}}$—Lame coefficients, where g_{kk} are the corresponding components of the metric tensor (fundamental object). Taking into account the Lame coefficients, the curvilinear components of acceleration W_k equal to zero have the form:

$$\sqrt{g_{kk}} \cdot W_k = \frac{d}{dt}\left(\frac{\partial v^2/2}{\partial \dot{q}^k}\right) - \frac{\partial v^2/2}{\partial q^k} = 0 \qquad (1)$$

Given the value of the square of velocity $v^2 = g_{ij}\dot{q}^i\dot{q}^j$, we obtain after the transformations:

$$\ddot{q}^m + \frac{1}{2}g^{mk}\left[\frac{\partial g_{kj}}{\partial q^i} + \frac{\partial g_{ik}}{\partial q^j} - \frac{\partial g_{ij}}{\partial q^k}\right]\dot{q}^i\dot{q}^j = 0 \qquad (2)$$

where $\tilde{A}_{ij}^m = \frac{1}{2}g^{mk}\left[\frac{\partial g_{kj}}{\partial q^i} + \frac{\partial g_{ik}}{\partial q^j} - \frac{\partial g_{ij}}{\partial q^k}\right]$ is a Christoffel 2nd type affinity factor (non-tensor-type).

Finally, the equation of optimal trajectories is:

$$\ddot{q}^m + \tilde{A}_{ij}^m \dot{q}^i\dot{q}^j = 0 \qquad (3)$$

$m = 1, 2, \dots, M$; M is the dimension of space. The system will consist of M second order differential equations and have $2M$ integration constants.

Next, we determine the Lame coefficients and analyze the applied problem for three factors.

The decision-making surface is approximated by the function of two variables of the form $z(x, y) = a_0 a_1^x a_2^y$ (approximation implies the possibility of using other types of function). The values of the parameters a_0, a_1, a_2 are determined from the results of statistical processing. We choose scale coefficients so that the analyzed values z, x, y are dimensionless. Let us introduce the generalized coordinates:

h—is the height of the points of the decision surface above the $x0y$ plane and the azimuth angle φ of the points of the decision surface, counterclockwise in a positive direction, that is, the generalized coordinates of the points—(h, φ).

After no complex transformations, in generalized coordinates, the radius vector of the points on the decision surface will look like:

$$\vec{R}(h, \phi) = \begin{bmatrix} \dfrac{\ln\left(\frac{h}{a_0}\right)Cos(\phi)}{Cos(\phi)\ln(a_1)+Sin(\phi)\ln(a_2)} \\ \dfrac{\ln\left(\frac{h}{a_0}\right)Sin(\phi)}{Cos(\phi)\ln(a_1)+Sin(\phi)\ln(a_2)} \\ h \end{bmatrix} = \begin{bmatrix} \dfrac{\ln\left(\frac{h}{a_0}\right)}{\ln(a_1)+tg(\phi)\ln(a_2)} \\ \dfrac{\ln\left(\frac{h}{a_0}\right)}{ctg(\phi)\ln(a_1)+\ln(a_2)} \\ h \end{bmatrix} \tag{4}$$

We introduce the notation $\frac{1}{a_0} = A_0$, $\ln(a_1) = A_1$, $\ln(a_2) = A_2$. Determine the local basis (benchmark):

$$\vec{g}_\phi = \begin{bmatrix} -\dfrac{A_2\ln(A_0h)}{(A_1Cos(\phi)+A_2Sin(\phi))^2} \\ \dfrac{A_1\ln(A_0h)}{(A_1Cos(\phi)+A_2Sin(\phi))^2} \\ 0 \end{bmatrix}; \quad \vec{g}_h = \begin{bmatrix} \dfrac{1}{(A_1+A_2tg(\phi))h} \\ \dfrac{1}{(A_2+A_1ctg(\phi))h} \\ 1 \end{bmatrix} \tag{5}$$

We introduce additional notations to simplify the calculations:

$$B(\phi) = A_2Sin(\phi) + A_1Cos(\phi);$$
$$M(\phi) = A_1 + A_2tg(\phi);$$
$$N(\phi) = A_2 + A_1ctg(\phi); \tag{6}$$

Given the notation (6), the metric tensor will have the form:

$$g_{ij} = \begin{bmatrix} \left(\dfrac{\ln(A_0h)}{B^2(\phi)}\right)^2\left(A_1^2 + A_2^2\right) & \dfrac{\ln(A_0h)}{B^2(\phi)h}\left(\dfrac{A_1}{N(\phi)} - \dfrac{A_2}{M(\phi)}\right) \\ \dfrac{\ln(A_0h)}{B^2(\phi)h}\left(\dfrac{A_1}{N(\phi)} - \dfrac{A_2}{M(\phi)}\right) & \dfrac{1}{h^2}\left(\dfrac{1}{M^2(\phi)} + \dfrac{1}{N^2(\phi)}\right)+1 \end{bmatrix} \tag{7}$$

To calculate the values of the symbols of the 2nd type of Christoffel, we find the determinant g of the metric tensor g_{ij} and the object of the upper structure g^{pi}:

$$g = \left(\dfrac{\ln(A_0h)}{B^2(\phi)}\right)^2\left(A_1^2 + A_2^2\right)$$
$$\times \left(\dfrac{1}{h^2}\left(\dfrac{1}{M^2(\phi)} + \dfrac{1}{N^2(\phi)}\right)+1\right) - \left(\dfrac{\ln(A_0h)}{B^2(\phi)h}\left(\dfrac{A_1}{N(\phi)} - \dfrac{A_2}{M(\phi)}\right)\right)^2 \tag{8}$$

$$g^{pi} = \begin{bmatrix} \left(\frac{1}{h^2}\left(\frac{1}{M^2(\phi)} + \frac{1}{N^2(\phi)}\right) + 1\right)\Big/_g & \left(-\frac{\ln(A_0 h)}{B^2(\phi)h}\left(\frac{A_1}{N(\phi)} - \frac{A_2}{M(\phi)}\right)\right)\Big/_g \\ \left(-\frac{\ln(A_0 h)}{B^2(\phi)h}\left(\frac{A_1}{N(\phi)} - \frac{A_2}{M(\phi)}\right)\right)\Big/_g & \left(\frac{\ln(A_0 h)}{B^2(\phi)}\right)^2\left(A_1^2 + A_2^2\right)\Big/_g \end{bmatrix} \tag{9}$$

Enter the designation of generalized coordinates:

$$q^1 = \phi, \quad q^2 = h \tag{10}$$

To reconcile the index notation, we rewrite (3) in the following representation:

$$\ddot{q}^p + \tilde{A}_{kl}^p \dot{q}^k \dot{q}^l = 0 \tag{11}$$

Symbol of the 2nd type of Christoffel:

$$\tilde{A}_{kl}^p = \frac{1}{2}g^{pi}\left[\frac{\partial g_{ik}}{\partial q^l} + \frac{\partial g_{il}}{\partial q^k} - \frac{\partial g_{kl}}{\partial q^i}\right] \tag{12}$$

In expanded form it will look like:

$$\tilde{A}_{kl}^p = \begin{pmatrix} \tilde{A}_{11}^1 & \tilde{A}_{21}^1 \\ \tilde{A}_{12}^1 & \tilde{A}_{22}^1 \\ \tilde{A}_{11}^2 & \tilde{A}_{21}^2 \\ \tilde{A}_{12}^2 & \tilde{A}_{22}^2 \end{pmatrix} \tag{13}$$

5.1 Interpretation of Results and Their Evaluation

To construct the decision-making surface, we will use the actual values of the parameters obtained in a 15-minute session of data transmission from the board of an artificial satellite of the Earth "Ocean—1". The parameters studied have a sufficient level of correlation. The first parameter under study is the current of solar cells, denote it as $z(t)$, the second parameter under test is the load current, denote it as $x(t)$, the third parameter under test is the load voltage, denote it as $y(t)$. All actual parameters for a session are listed in Table 1.

Using numerical methods we obtain an approximate equation of the decision surface:

$$z(x, y) = 0,00645 \cdot (23,976)^x \cdot (17,644)^y \tag{14}$$

A fragment of the matrix (size 30×30) with relative values $z(x, y) = S$ is shown in the Fig. 3:

Table 1 The values of the investigated parameters per communication session

Session time, min	Current of solar panels (TC-1), z(t), A		Load current (TH-1) x(t), A		Load voltage (HH-1) y(t), V	
	Range ($A_{max} = 21$ A)		Range ($A_{max} = 50$ A)		Range ($U_{max} = 34$ V)	
	A	%/100	A	%/100	U	%/100
0	6.69	0.318571	5.77	0.1154	32.12	0.944706
0.5	6.09	0.29	10.08	0.2016	31.8	0.935294
1	1.28	0.060952	5.12	0.1024	31.59	0.929118
1.5	0.57	0.027143	6.56	0.1312	33	0.970703
2	4.24	0.201987	9.93	0.198586	31.59	0.929171
2.5	1.86	0.088792	8.87	0.177369	32.2	0.947124
3	5.11	0.243274	11.25	0.2251	32.18	0.946418
3.5	2.37	0.112661	14.7	0.293948	31.22	0.918111
4	1.63	0.077577	10.13	0.202648	32.79	0.964451
4.5	3.04	0.14472	11.53	0.230511	31.58	0.928816
5	2	0.095192	10.42	0.208421	32.36	0.951876
5.5	2.99	0.142477	12.64	0.25275	30.59	0.899635
6	3.4	0.162056	10.7	0.213901	32.06	0.943068
6.5	4.29	0.20443	12.43	0.248514	32.31	0.950285
7	2.95	0.140634	4.76	0.095147	32.85	0.966163
7.5	5.8	0.276256	13.42	0.268416	30.25	0.88965
8	1.63	0.077591	7.94	0.158792	32.17	0.946271
8.5	4.94	0.235122	4.17	0.083367	30.39	0.893866
9	5.89	0.280563	13.38	0.267676	33.18	0.975891
9.5	6.25	0.297733	11.32	0.226454	31.8	0.935411
10	2.83	0.134952	6.31	0.126161	30.35	0.892605
10.5	3.76	0.178893	6.97	0.139363	32.59	0.958582
11	3.96	0.188503	6.63	0.132647	32.4	0.953046
11.5	4.68	0.222879	6.04	0.120858	30.42	0.894815
12	3.27	0.155753	12.38	0.247658	30.37	0.893201
12.5	4.23	0.201609	13.67	0.273471	32.33	0.950742
13	4.43	0.210772	8.38	0.16762	32.66	0.960679
13.5	6.39	0.304152	12.83	0.256608	32.21	0.947246
14	6.61	0.314822	12.05	0.240982	32.76	0.963529
14.5	2.29	0.108869	8.09	0.161817	30.89	0.908413
15	4.22	0.200917	7.07	0.141479	31.66	0.931191

	0	1	2	3	4	5	6	7	8
0	0.118	0.128	0.106	0.114	0.125	0.134	0.121	0.13	0.12
1	0.135	0.146	0.122	0.131	0.143	0.153	0.139	0.149	0.138
2	0.177	0.193	0.16	0.172	0.188	0.202	0.183	0.196	0.181
3	0.15	0.164	0.136	0.146	0.16	0.171	0.155	0.167	0.154
4	0.21	0.228	0.189	0.204	0.222	0.238	0.216	0.232	0.214
5	0.133	0.145	0.12	0.129	0.141	0.151	0.137	0.147	0.136
6	0.194	0.211	0.175	0.188	0.205	0.22	0.2	0.215	0.198
7	0.146	0.158	0.131	0.142	0.154	0.166	0.15	0.161	0.149
8	0.125	0.136	0.113	0.122	0.133	0.143	0.129	0.139	0.128
9	0.13	0.142	0.118	0.127	0.138	0.148	0.134	0.144	0.133
10	0.235	0.256	0.213	0.229	0.25	0.268	0.243	0.261	0.241
11	0.128	0.139	0.115	0.124	0.136	0.145	0.132	0.142	0.131
12	0.118	0.129	0.107	0.115	0.126	0.135	0.122	0.131	0.121
13	0.171	0.186	0.154	0.166	0.181	0.194	0.176	0.189	0.175
14	0.179	0.195	0.162	0.175	0.19	0.204	0.185	0.199	0.184
15	0.132	0.144	0.119	0.129	0.14	0.15	0.136	0.146	...

$S =$ (label to the left of the table, at row 7)

Fig. 3 A fragment of the matrix (size 30×30) with relative values $z(x, y) = S$

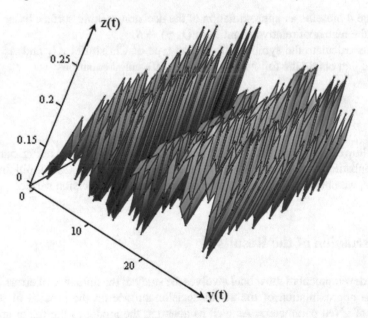

Fig. 4 Approximation of the decision surface $z(x, y) = S$

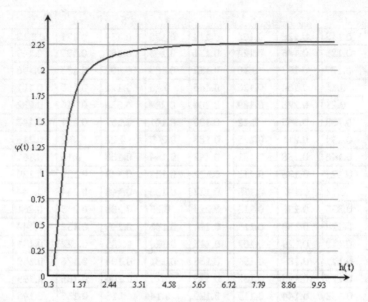

Fig. 5 The equation of the optimal trajectory on the decision surface $z(x, y) = S$

Figure 4 presents an approximation of the decision-making surface based on the data of the matrix of relative quantities $z(x, y) = S$.

Let us calculate the symbols of the 2nd type of Christoffel $\tilde{A}_{11}^1$ and $\tilde{A}_{21}^2$ using Mathcad and obtain the following system of differential equations:

$$\begin{cases} \ddot{q}^1 + \tilde{A}_{11}^1 (\dot{q}^1)^{(2)} = 0 \\ \ddot{q}^2 + \tilde{A}_{21}^2 \dot{q}^2 \dot{q}^1 = 0 \end{cases} \tag{15}$$

which allowed us to construct the required equation $\varphi(h)$ presented in Fig. 5, respectively. Substituting into the system (11) the other values of the Christoffel 2nd type symbols, we obtain variants of optimal trajectories on the decision surface.

6 Discussion of the Results

Further development of the model involves the study of the influence of errors arising from the approximation of the actual decision surface on the process of optimal control of given parameters. As well as adapting the model to the factor space of higher dimensions.

7 Conclusions

The proposed model allows optimal control of the set parameters of technical systems in order to prevent emergencies and risk situations. Based on this model, it is possible to create decision support systems for organizing activities aimed at reducing risk situations in technical systems.

References

1. Popov, O., Iatsyshyn, A., Kovach, V., Artemchuk, V., Taraduda, D., Sobyna, V., Sokolov, D., Dement, M., Yatsyshyn, T., Matvieieva, I.: Analysis of possible causes of npp emergencies to minimize risk of their occurrence. Nucl. Radiati. Saf. **1**(81), 75–80 (2019). https://doi.org/10.32918/nrs.2019.1(81).13
2. Popov, O.O., Iatsyshyn, A.V., Kovach, V.O., Artemchuk, V.O., Kameneva, I.P., Taraduda, D.V., Sobyna, V.O., Sokolov, D.L., Dement, M.O., Yatsyshyn, T.M.: Risk assessment for the population of Kyiv, Ukraine as a result of atmospheric air pollution. J. Health Poll. **10**(25), 200303 (2020). https://doi.org/10.5696/2156-9614-10.25.200303
3. Popov, O., Iatsyshyn, A., Kovach, V., Artemchuk, V., Taraduda, D., Sobyna, V., Sokolov, D., Dement, M., Hurkovskyi, V., Nikolaiev, K., Yatsyshyn, T., Dimitriieva, D.: Physical features of pollutants spread in the air during the emergency at NPPs. Nucl. Radiat. Saf. **4**(84), 88–98 (2019). https://doi.org/10.32918/nrs.2019.4(84).11
4. What is Remote Sensing? The Definitive Guide. Access mode: https://gisgeography.com/remote-sensing-earth-observation-guide/
5. Shmelova, T., Sikirda, Y., Scarponi, C., Chialastri, A.: Deterministic and stochastic models of decision making in air navigation socio-technical system. In: CEUR Workshop Proceedings, vol. 2104, pp. 649–656 (2018). http://ceur-ws.org/Vol-2104/paper_221.pdf
6. Grishin, I., Timirgaleeva, R.: Air navigation: automation method for controlling the process of detecting aircraft by a radar complex. In 2019 24th Conference of Open Innovations Association, pp. 110–115 (2019). https://doi.org/10.23919/fruct.2019.8711905
7. Kondratyeva, N., Valeev, S.: Fatigue test optimization for complex technical system on the basis of lifecycle modeling and big data concept. In 2016 IEEE 10th International Conference on Application of Information and Communication Technologies, pp. 1–4 (2016). https://doi.org/10.1109/icaict.2016.7991656
8. Leveson, N.: A new accident model for engineering safer systems. Saf. Sci. **42**(4), 237–270 (2004)
9. Raza, A.: Mathematical model of corrective maintenance based on operability checks for safety critical systems. Am. J. Applied Math. **6**(1), 8–14 (2018)
10. Mokhor, V., Gonchar, S., Dybach, O.: Methods for the total risk assessment of cybersecurity of critical infrastructure facilities. Nucl. Radiat. Saf. **2**(82), 4–8 (2019). https://doi.org/10.32918/nrs.2019.2(82).01
11. Bilan, T., Rezvik, I., Sakhno, O., But, O., Bogdanov, S.: Main approaches to cable aging management at nuclear power plants in Ukraine. Nucl. Radiat. Saf. **4**(84), 54–62 (2019). https://doi.org/10.32918/nrs.2019.4(84).07
12. Mokhor, V.V., Tsurkan, O.V., Tsurkan, V.V., Herasymov, R.P.: Information security assessment of computer systems by socio-engineering approach. In: Selected Papers of the XVII International Scientific and Practical Conference on Information Technologies and Security (ITS 2017), November 30, 2017, pp. 92–98. Kyiv, Ukraine (2017). http://ceur-ws.org/Vol-2067/paper13.pdf

13. Yatsyshyn, T., Shkitsa, L., Popov, O., Liakh, M.: Development of mathematical models of gas leakage and its propagation in atmospheric air at an emergency gas well gushing. East. Eur. J. Enter. Technol. **5/10**(101), 49–59 (2019). https://doi.org/10.15587/1729-4061.2019.179097

14. Zhang, S., Du, M., Tong, J., Li, Y.-F.: Multi-objective optimization of maintenance program in multi-unit nuclear power plant sites. Reliab. Eng. Sys. Saf. (2019). https://doi.org/10.1016/j.ress.2019.03.034

15. Chen, Z.C., Liu, P.H., Pei, Z.: An approach to multiple attribute group decision making based on linguistic intuitionistic fuzzy numbers. Int. J. Comput. Int. Sys. **8**, 747–760 (2015)

16. Tang, J., Meng, F., Zhang, Y.: Decision making with interval-valued intuitionistic fuzzy preference relations based on additive consistency analysis. Inf. Sci. **467**, 115–134 (2018). https://doi.org/10.1016/j.ins.2018.07.036

17. Kovach, V., Lysychenko, G. Toxic soil contamination and its mitigation in Ukraine. In: Dent, D., Dmytruk, Y. (eds.) Soil Science Working for a Living. Springer, Cham (2017). https://doi.org/10.1007/978-3-319-45417-7_18

18. Gomez Fernandez, M., Tokuhiro, A., Welter, K., Wu, Q.: Nuclear energy system's behavior and decision making using machine learning. Nucl. Eng. Des. **324**, 27–34 (2017). https://doi.org/10.1016/j.nucengdes.2017.08.020

19. Reyes, J.N., Groome, J., Woods, B.G., Young, E., Abel, K., Yao, Y., Yoo, Y.J.: Testing of the multi-application small light water reactor (MASLWR) passive safety systems. Nucl. Eng. Des. **237**(18), 1999–2005

20. Popov, O., Iatsyshyn, A., Kovach, V., Artemchuk, V., Taraduda, D., Sobyna, V., Sokolov, D., Dement, M., Yatsyshyn, T.: Conceptual approaches for development of informational and analytical expert system for assessing the NPP impact on the environment. Nucl. Radiat. Saf. **3**(79), 56–65 (2018). https://doi.org/10.32918/nrs.2018.3(79).09

21. Wang, D., Gu, X., Zhou, G., Li, S., Liang, H.: Decision-making optimization of power system extended black-start coordinating unit restoration with load restoration. Int. Trans. Elec. Energy Sys. **27**(9), e2367 (2007). https://doi.org/10.1002/etep.2367

Problems, Methods and Means of Monitoring Power Losses in Overhead Transmission Lines

Ihor Blinov⊙, Ievgev O. Zaitsev⊙, and Vladislav V. Kuchanskyy⊙

Abstract The problems of capacity allocation of cross-border transmissions for electricity market coupling are considered. Ways to improve transmission efficiency electricity in 330–750 kV bulk electrical power networks are considered. The transmission efficiency is reduced due to the technological losses in the electricity elements. The main technological losses are load losses in the elements of the network and losses on the corona in the wires overhead lines. It is proposed to control in real time load losses and power losses on the crown in the main air transmission lines. Power losses to the corona are significantly dependent on the weather conditions and voltage level in the network. A slight decrease in the voltage in the grid during rain and when deposited on the frost wires will significantly reduce the loss of electrical energy in the wires of extra-voltage overhead lines. The aim of the work is based on the project "Energy Bridge" Ukraine—"European Union", which envisages the launch in Ukraine of 750 kV transmission lines "Khmelnitsky NPP—Rzeszow" and "South Ukrainian NPP-Isakcha". Improving the reliability and efficiency of interconnection will ensure the creation of conditions for Ukraine's energy independence by expanding the use of its energy potential and throughput of the Ukrainian grid, fulfilling the main tasks of Ukraine's energy development.

Keywords Corona discharge · Crucial electrical density · Electrical power losses · Optical information measuring systems · Control accuracy · Smart grid · Controlled shunt reactors · Energy bridge · Decentralized market coupling · Flow-based market coupling · FACTS

I. Blinov · I. O. Zaitsev · V. V. Kuchanskyy (✉)
Institute of Electrodynamics National Academy of Sciences of Ukraine, Kyiv, Ukraine
e-mail: kuchanskiyvladislav@gmail.com

V. Babak et al. (eds.), *Systems, Decision and Control in Energy I*, Studies in Systems, Decision and Control 298, https://doi.org/10.1007/978-3-030-48583-2_8

1 Capacity Allocation of Cross-Border Transmissions for Electricity Market Coupling

Ukraine has assumed international obligations on the implementation of European legislation and adopted a Law introducing a new electricity market in Ukraine [1, 2]. The market reforms defined in the Law in the electric power industry can both different affect the further development of the Ukrainian electricity sector.

This will have a corresponding impact on all sectors of the Ukrainian economy, as the assumptions of errors at each stage of the reforms in this area have a great price for the country as a whole [3].

Since 2019, a new market model has been implemented in Ukraine. New roles of participants and new rules of their work are defined [4]. An important advantage of the new model of a liberalized electricity market is the ability to import and export electricity.

First, there is an opportunity to electricity markets coupling of Ukraine and 4MC countries (Hungary, Romania, Slovakia and the Czech Republic). There are also opportunities to implement projects for the export of electricity by nuclear power plants (NPP) to Poland. An example of such a project "Energy Bridge" Ukraine—"European Union", which envisages the launch in Ukraine of 750 kV transmission lines "Khmelnitsky NPP—Rzeszow" and "South Ukrainian NPP-Isakcha".

The interconnected power system (IPS) of Ukraine may be used effectively as a transit node in the electricity exchange between other countries. The method of analysis of bids on capacity for several cross-border transmissions must provide the equal energy exchange between two power systems through IPS of Ukraine as transit node.

The forming of coupled cross-border capacity market demands the solving of two main tasks. The first one is the development of method for trading on separate cross-border transmission taking into account the bidirectional electricity exchange. The second task is the development of method of exchange balancing on several cross-border transmissions taking into account the power flows between two markets, which transit through third market.

One of the most important and actual tasks of operability assurance of day-ahead market [5] is to determine the cross-border congestion management methods. To date there are two main approaches to solving this problem. These are the market splitting and the market coupling [6].

The market-coupling approach is more promising and actively developing now. The most known implementations of market coupling are the flow-based market coupling (FBC) and decentralized market coupling (DMC) [7, 8].

Therefore, market-coupling method is more effective for the developing of a typical method of solving the problem of cross-border congestion management.

The requirements for the market coupling and congestion management method are formulated. The main ones are:

- solving the problem of cross-border congestion management should be implemented by market coupling method, i.e. common electricity market previously split into local bidding areas, which are interconnected with potentially
  problematic cross-border interconnections;
- estimation of cross-border exchange volumes between the biding areas and
  calculation of clearing prices should be divided into separate sub-tasks;
- cross-border congestion management method must allow optimization by discrete
  variables.

The market coupling process requires the definition of transmission capacity of
intersystem (international) electric networks. It depends on the conditions, limitations
of the operation and losses of electric lines of high voltage (HV) and extra-high
voltage (EHV), including the corona losses.

2 The Problems of Calculating the Corona Losses of Overhead Lines

One of the significant problems of the electric power industry is the reduction of
power and electricity losses in a united energy system. The solution to this problem
is possible by optimizing the operating mode of the power system according to voltage
levels and reactive power flows. To solve this problem, it is necessary to develop a
system for measuring the operating modes of an extra-high voltage network, taking
into account the power losses per corona [9–19].

Electricity losses in overhead power lines of power transmission consist of load
losses, losses on the corona and losses from leakage currents through insulation, the
determining ones are load losses and losses on the corona, which to different degrees
depend on the voltage level: load losses at a constant value power, resistance, and
hence the voltage on the load side, are inversely proportional to the square of the
line voltage, and the crown loss is proportional to the line voltage to the fifth degree.
Thus, the optimal voltage level in the nodes of the power system depends on the ratio
of the losses per crown and load losses of the overhead lines. If in good weather,
load losses prevail over losses on the crown, then in bad weather (snow, rain, frost),
losses on the crown increase by 1–2 orders of magnitude [14–18].

The design of the overhead line (OL) of high and extra-high voltage is determined
by comparing construction costs with the cost of electricity losses in competing
options. If the energy losses are determined incorrectly, then the cross section of the
line may turn out to be either greater or less than optimal. If the estimated losses of
electricity to construction are underestimated, a variant with reduced phase conductor
cross-sections will be proposed. In this case, the overhead line parameters are not
optimal and during operation of such a line there will be increased energy losses.
Therefore, obtaining reliable calculated values of electric power losses in overhead
lines is extremely important. If the energy loss in the OL wires can be determined
by knowing the expected line load, then estimating the energy loss to the corona

requires not only knowledge of the meteorological situation at the OL passage, but also the use of calculation methods that give close to reality results [11, 12].

The power loss per crown in the OL depends on the parameters of weather conditions, the variations of which are random. Since the characteristics of the variability of power losses from the corona depend on the type of weather, it is usually considered separately the power losses per corona in good weather, good weather with high humidity, dry and wet snow, fog, rain, frost. Since the recurrence of meteorological phenomena is determined by the period of solar activity within which the loss of the corona can be considered a stationary random process the characteristics of which are its distribution function and correlation function. Therefore, to study the characteristics of corona losses, it is necessary to have data on continuous measurements of weather parameters and corona losses during the period of solar activity.

Today, outdated methods are used in bulk electric networks, which underestimate both specific power losses per corona and annual electric power losses per corona, as defined in [15–19]. The mentioned materials create the illusion of a favorable situation with losses in high and extra voltage overhead lines. In fact, a large amount of electricity is lost in the backbone lines, and the maximum power loss per crown consumes a tangible fraction of the installed capacity of power plant generators. The use of an underestimating energy loss per crown technique when designing new overhead lines multiplies the number of economically inefficient lines introduced.

That is why the creation of a continuous measurement system for the overhead corona losses is a necessary basis for optimizing the voltage and reactive power flows. Accounting for corona losses can have a significant economic effect in light of the changed principles for calculating tariffs for electric power transmission and distribution services, as well as international programs for energy conservation and reduction of environmental impact on electricity production and transmission that have been carried out in recent years.

3 The Reduction of Energy Loss upon Implementation Smart Grid Conception

Today, the development of electric networks takes into account the Smart Grid concept [21–23], an important part of which is FACTS systems. On the basis of modern power electronics, effective FACTS devices are developed for flexible control of modes of power systems. FACTS are capable of simultaneously affecting voltage, reactance and voltage angle. It is known that the calculations of steady-state modes of power systems are the most frequently performed tasks at all territorial and temporal levels of control and planning of modes. FACTS include devices that are designed to stabilize the voltage, increase stability, optimize the distribution of power flows, and reduce losses in electrical networks. One of the typical examples of FACTS devices are controlled shunt reactors (CSR) [24–34], which in power systems perform a wide range of tasks, one of which is to increase throughput and reduce power losses. The above device solves the problem of converting the electrical network from a passive element of the transmission of electricity into active, which provides control of the modes of operation of the bulk electrical network [28–30].

The use of non-phase modes of operation of traditional uncontrolled shunt reactors has certain limitations that make it impossible to use them at substations located at NPP [20, 34–40]. These limitations are the possibility of switching overvoltages with the development of a severe systemic accident. On the other hand, there is a possibility of self-excitation of generators at the NPP in the case of low-load or idling of the line of transmission of high-voltage. To ensure reliable operation of the backbone electrical network, it is necessary to use FACTS technical devices in the form of CHR, which allow regulating the power flows in the modes of operation of the lines of high-voltage transmission.

It should be noted that a change in the inductance of the controlled shunt reactor will reduce the value of power losses, thereby increasing the efficiency of the over-voltage transmission line [15, 16]. The transmission of power through the transmission lines of higher classes of voltage encounters a number of serious technical problems. One such problem is the reduction of the efficiency in the transmission of low power. We show that in high-voltage power lines with controlled and unmanaged shunt reactors, it is possible to significantly increase the efficiency of power transmission in modes of loading of less natural power.

The transmission of power through the transmission lines of higher classes of voltage encounters a number of serious technical problems. One such problem is the reduction of the efficiency in the transmission of low power. We will show that in high-voltage transmission lines with controlled and unmanaged shunt reactors, it is possible to significantly increase the transmission efficiency in modes of loading of less natural power.

$$
\begin{aligned}
\Delta P = {} & G \cdot (B^2 \cdot A1 - 6 \cdot (A2 + A3) + G^2 \cdot (R^2 + X) + 8) \cdot U^2 \\
& + \left(\frac{G \cdot A1}{2} + R \right) \cdot \frac{P^2 + Q^2}{U^2} + \left(A3 \cdot (3 \cdot A2 - 4(A3 - 4)) - \frac{A4^2 + G^2 \cdot X^2}{2} \right) \cdot P \\
& + \left(\frac{A4 \cdot A2 - A4 \cdot A3 - 7 \cdot G \cdot A2}{2} - A4 + 15 \cdot G \cdot X \right) \cdot Q
\end{aligned}
\tag{1}
$$

where R—active resistance, Ohm; X—inductive resistance, Ohm; G—active conductivity, S; B—reactive conductivity, S; U—rated voltage, kV; P—active power, MW; P—reactive power, MWAr. For the sake of convenience, additional coefficients have been applied: $A1 = R^2 + X^2$, $BX = A2$, $GR = A3$, $BR = A4$.

The efficiency of application such device of FACTS as controlled shunt reactor shown on Fig. 1. As we can see, value of active power losses ΔP are widely varied depending on value of inductance of controlled shunt reactors. It should be noted, that in (1) overall losses consist of technological and corona losses. In case of application controlled shunt reactors occurs reducing technological losses and consequently overall losses. At smooth control of controlled shunt reactors we have to find the value of inductance of controlled shunt reactor at which the overall losses will be minimum.

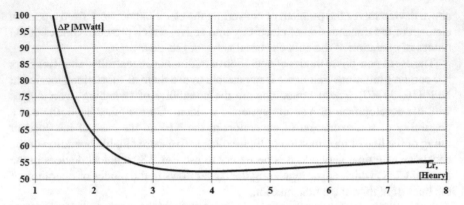

Fig. 1 Application efficiency of controlled shunt reactors

4 Requirements for Corona Discharge Parameters Measurement System

Tension is an important fundamental force characteristic of EP [41]. Unlike other electrical quantities, the intensity of low-frequency electromagnets are unsupported by both methods and measuring instruments. In this regard, the means of measuring the intensity of low-frequency electronic devices fell out of the range of devices for measuring electrical quantities. This situation is especially typical for measuring instruments for industrial frequency electric drives on the surface of high-voltage insulators and near power lines.

The algorithm for measuring power losses per crown is based on the use of telemetric information about the operating parameters of the line issued by the system. According to the algorithm for calculation used the values of the current power losses, which are determined by subtracting the active power P2 at its end from the active power P1 entering the line at its end, and this compensates for the systematic and random error of the loss measurement.

To implement the measurement algorithm it was necessary to solve the following tasks were solved:

1. Consideration of the components of electricity losses in the EPS and existing methods for their calculation.
2. Analysis of the sensitivity of the method when measuring small values of electric energy losses on the corona (at high humidity) and in wires in real time.
3. The study of methodological and instrumental errors embedded in the software package algorithms for determining load losses and crown losses in real time.
4. Analysis of the effect of reducing electricity losses on the corona and in the wires of the power lines when regulating voltage according to data on current values of electricity losses.
5. Determination of the maximum voltage control range and the possible effect of saving power losses on a unit 500, 750 kV.

The considered algorithm is used in the measurement systems software package, designed to estimate active, reactive power, voltages and linear losses, and allows you to quickly monitor the level of crown loss to quickly take steps to reduce them. The measurement systems are also designed for:

- the definition of the class of corona discharge power losses in the power transmission lines;
- selection of diagnostic signals available for measurement, and control points on the object under study;
- development of a mathematical model of the diagnostic object, the analysis of which allows substantiating possible diagnostic parameters;
- development of algorithms for obtaining numerical values of selected diagnostic parameters;
- construction of decisive rules for identifying and classifying defects; creation of means implementing certain steps of the diagnostic process from the selected measurement and diagnostic signals before making diagnostic solutions.

The system software (primary data acquisition and processing module) consists of the following parts: data processing software for measurement module, a micro-controller configuration module for information-measuring channel, a primary data processing module. The specially designed software is used for the control system for data collection, processing, and analysis. This software is used for the operation control of primary sensor of corona discharge power losses measurement systems, to organize other hardware and software modules, for the primary process implementation, data collection and data transmission from the transducers to designed devices for the secondary processing, for the statistical data analysis, storage and control data display in a simple form for the operator.

Operations of calculating the value of corona discharge power losses special software solution is used. In turn, obtained at the work of the module of mathematical processing and module of automatic control of the state of the electrical equipment node is transferred to the data storage organization module for database management based on the history of measurements. In this case, it is possible to create knowledge bases with diagnostic features, which depend on the value of the physical parameter that is control of the certain state of the power equipment.

On Figs. 2 and 3 shown capacitance and electrical field strength of corona discharge.

5   Energy Losses Information Measuring Systems with Optical Sensing Elements

The problem of contactless measurements of electric field strength is rather urgent. Such measurements are necessary for the remote monitoring of losses in overhead power lines [42], including transmission line parameters such as discharge to the corona.

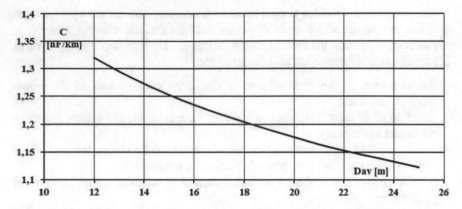

Fig. 2 Changes in capacitance depending on geometric mean distance

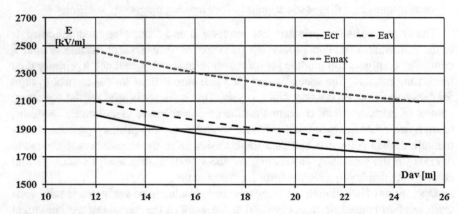

Fig. 3 Changes in the electric field depending on the geometric mean phase distance of the line

In this case, the means of losses in overhead power lines discharge are in a rather harsh environment of the operating environment, which is characterized by a significant intensity of electromagnetic fields, high temperature and humidity and ozone content.

In this case, the actual task is to increase the noise immunity of the monitoring and measurement of the losses in overhead power lines. The solution of this problem with the use of traditional structures discussed above today requires the use of sophisticated measuring equipment directly on the transmission line, so there is a need to implement measuring converters with the application of fundamentally new solutions, methods and methods of construction.

Perspective way to design of measurement system it is to use optical measurement system (OM) are used [43, 44]. This system usually consisting of optical sensing element for measurement and analyzer unit of parameters of line, they are completely inert to the effects of the fields of the internal environment of the powerful generators, although they have a higher cost of means being placed in the "safe zone" where there is no effect of the transmission line operating environment.

An essential feature of OM is their important and significant perspective in application, especially since, in addition to noise immunity, they operate in powerful electromagnetic fields; they also have considerable measurement distance (up to hundreds of meters), as well as high potential accuracy and measurement speed through the use of high-speed optical and resolution optical elements, despite the fact that the technology for the production of primary measurement converters for OM is currently sufficiently high complicated and expensive, requires the use of special technological and metrological equipment. So, in this case one of the necessary conditions it is that the optical sensing element or primary transducer of the sensor introduced a minimum of distortion in the structure of the field distribution, be passive (not requiring availability power supply in the measurement area), miniature and provided galvanic discharge between the measurement area and the equipment. These conditions are satisfied by fiber-optic sensors (FOS) based on the linear Pockels electro-optic effect. In this case, as sensitive elements can be used ferroelectric crystals of 3 m class.

On Fig. 4 shown block diagram design of the corona discharge power losses measurement systems with optical sensor network in the transmission lines of the high- and extra-high voltage transmission lines. The principle of operation of optical

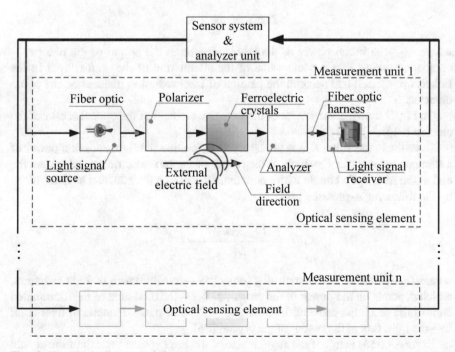

Fig. 4 The block diagram of the block diagram of corona discharge power losses optical measurement systems

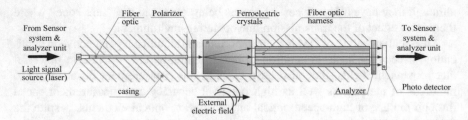

Fig. 5 Principle of the work of the optical sensing element

measuring systems is based on electro-optical modulation of a light wave due to the conversion of linearly polarized light to elliptically polarize. In this case, the longitudinal Pockels effect is used when the electric field is directed into the crystal along an axis parallel to the direction of light propagation.

To determine the function of converting of the one optical sensor (Fig. 5) from system of optical sensor network, let us consider the simplest case of using the specified design for optical fiber meters with amplitude modulation of the information signal. In this case, to determine the sensor conversion functions $L(\Phi)$, take advantage of the dependence of optical radiation in the sensor on the value of the attenuation of the optical light in the Pokels optical cell. So, $L(\Phi)$ calculated as

$$L(\Phi) = \Phi_0 \cdot K \tag{2}$$

where Φ_0—the initial power of the laser radiation at the output of the fiber; K—coefficient taking into account value of the attenuation of the optical light in the Pokels optical cell and account the amount of laser radiation incident on the photo detector.

On Fig. 5 shown construction of the sensitive element of the optical sensor of the electric field.

To evaluate the value K it is necessary to determine the light emission power of a semiconductor laser F, which, at the output of the ferroelectric crystal, enters the end of the fiber optic bundle with a core diameter of one fiber 2a, can be determined by the following expression [44]:

$$K_{Diff} \approx \frac{\left(\sum S_{OF}\right)^2}{L_1^2} \cdot \frac{1}{\pi} \cdot \Phi \cdot \chi. \tag{3}$$

where L_1—distance to the controlled object; χ—transmittance of light radiation, which depends on the power of the crown; ΣS_{OF}—the total area of the illuminated ends of the receiving cable of the fibers; $2a = 25$–$100\ \mu$m—diameter of fiber core; Φ—the light flux at the output of the crystal [26].

Expression (3) is true if the angular size of the laser spot at the exit of the crystal is less than the viewing angle of the receiving device [26].

The coefficient taking into account value of the attenuation of the optical light in the Pokels optical cell calculated as the intensity Φ of the light flux at the output of the crystal by next expression [45].

$$\Phi = \frac{1}{2}\Phi_0\left[1 + \left\{\frac{2\pi}{\lambda}\cdot n_0^3 r_{41}\cdot E_i\frac{\sin(\Theta\cdot l)}{\Theta}\right\}\times \times \sin(2\varphi - \Theta\cdot l)\right] \tag{4}$$

where Φ_0- light intensity at the input of the crystal; $\lambda-$ wavelength of monochromatic light; n_0- refractive index; E_i- electric field strength in crystal; $\Theta-$ optical activity of the electro-optical crystal; $l-$ length of the crystal; $r_{41}-$ electro-optical coefficient.

Get (3) and value Φ_0 expressions for transformation function on the out of system (2), can be written as

$$L(P) = \Phi_0\cdot\frac{1}{\pi}\cdot\chi\cdot\frac{\left(\sum S_{OF}\right)^2}{L_1^2}\cdot\left[1 + \left\{\frac{2\pi}{\lambda}\cdot n_0^3 r_{41}\cdot E_i\frac{\sin(\Theta\cdot l)}{\Theta}\right\} \times \sin(2\varphi - \Theta\cdot l)\right] \tag{5}$$

Using turn (5), it is possible to obtain the power value at the measurement points by the system in high and ultra-high voltage power lines. Given the obtained values, the system calculates power losses. The measurement results allow one to analytically determine energy losses (load losses, losses on the corona with leakage currents) in real time, which allows optimizing the operation mode of the overhead line in order to reduce these losses and achieve significant energy savings.

References

1. Ivanov, H.A., Blinov, I.V., Parus, Ye.V.: Imitation modeling of the balancing electricity market functioning taking into account system constraints on the parameters of the IPS of Ukraine mode. Tekhnichna Elektrodynamika **6**, 72–79, (Ukr) (2017). https://doi.org/10.15407/techned2017.06.072
2. Zhuikov, V., Pichkalov, I., Boyko, I., Blinov, I.: Price formation in the energy markets of Ukraine. In: Electronics and Nanotechnology (ELNANO), 2015 IEEE 35th International Conference. https://doi.org/10.1109/elnano.2015.7146953
3. Ivanov, H., Blinov, I., Parus Ye.: Simulation model of new electricity market in Ukraine. In: IEEE 6th International Conference on Energy Smart Systems (ESS), pp. 339–343 (2019). https://doi.org/10.1109/ess.2019.8764184
4. Blinov, I., Tankevych, S.: The harmonized role model of electricity market in Ukraine. In: Intelligent Energy and Power Systems (IEPS), 2016 2nd International Conference on (2016). https://doi.org/10.1109/ieps.2016.7521861
5. Kyrylenko, O.V., Blinov, I.V., Parus, Ye.V., Ivanov, H.A.: Simulation model of day ahead market with implicit considerarion of power systems network constraints. Tekhnichna Elektrodynamika **5**, 60–67, (Ukr) (2019). https://doi.org/10.15407/techned2019.05.060
6. Blinov, I.V.: New approach to congestion management for decentralized market coupling using net export curves. CIGRE Session 46 Paris 2016. Water and Energy Int. **61**(5), p. 76 (2018)

7. Blinov, I.V., Parus, Y.V.: Features of use of the net export function properties for the congestion management on the "day-ahead" market. Technical Electrodynamics **6,** pp. 63–68, (Ukr) (2015)
8. Blinov, I.V., Parus, Y.V.: Congestion management and minimization of price difference between coupled electricity markets. Technical Electrodynamics **4**, pp. 81 –88, (Ukr) (2015)
9. Kuznetsov, V., Tugay, Y., Kuchanskyy, V.: Influence of corona discharge on the internal ovevoltages in highway electrical networks. Techn. Electrodyn. **6**, 55–60 (2017)
10. Tamazov, A.I.: Losses on the crown in high-voltage overhead power lines, vol. 571. Sputnik + , Moscow (2016). ISBN 978-5-9973-3869-5
11. Riba, Jordi-Roger, Abomailek, Carlos, Casals-Torrens, Pau, Capelli, Francesca: Simplification and cost reduction of visual corona test. Gen. Trans. Distrib. IET **12**(4), 834–841 (2018). https://doi.org/10.1049/iet-gtd.2017.0688
12. Wang, Jialong, Vue, Bo, Deng, Xiguo, Liu, Teqing, Peng, Zongren: Electric field evaluation and optimization of shielding electrodes for high voltage apparatus in $\pm$ 1100 kV indoor DC yard. Dielectr. Electric. Ins. IEEE Trans. **25**(1), 321–329 (2018). https://doi.org/10.1109/TDEI. 2018.006890
13. Hernandez-Guiteras, J., Riba, J., Casals-Torrens, P.: Determination of the corona inception voltage in an extra high voltage substation connector. IEEE Trans. Dielectr. Electr. Insul. **20**(1), 82–88 (2013). https://doi.org/10.1109/TDEI.2013.6451344
14. Liu, Y.-P., You, S.-H., Wan, Q.-F., Chen, W.-J.: Design and realization of AC UHV corona loss monitoring system. High Voltage Eng **39**(9), 1797–1801 (2008)
15. Lu, F.-C., You, S.-H., Liu, Y.-P., Wan, Q.-F., Zhao, Z.-B.: AC conductors' corona-loss calculation and analysis in corona cage. IEEE Trans. Power Del. **27**(2), 877–885, Apr. (2012). https://doi.org/10.1109/TPWRD.2012.2183681
16. Liu, Y., Huang, S., Liu, S., Liu, D.: A helical charge simulation based 3-D calculation model for corona loss of AC stranded conductors in the corona cage Aip Advances, **8**(1), 015303 (2018). https://doi.org/10.1063/1.5017244
17. Suleimanov, V.N., Katsadze, T.L.: Electric networks and systems—K.: NTUU "KPI", p. 504 (2007). ISBN 978-966-622-246-9
18. Liu, Yunpeng, Chen, Sijia, Huang, Shilong: Evaluation of Corona Loss in 750 kV Four-Circuit Transmission Lines on the same tower considering complex meteorological conditions. IEEE Access **6**, 67427–67433 (2018). https://doi.org/10.1109/ACCESS.2018.2878763
19. Fanghui, Y., Farzaneh, M., Jiang, X.: Corona investigation of an energized conductor under various weather conditions. IEEE Trans. Dielect. Elec. Insul. **24**(1), 462–470 (February 2017). https://doi.org/10.1109/TDEI.2016.006302
20. Kuznetsov, V., Tugay, Y., Kuchanskyy, V.: Investigation of transposition EHV transmission lines on abnormal overvoltages. Techn. Electrodyn. **6**, 51–56 (2013)
21. Blynov, Y.V., Denysyuk, S.P., Zhuykov, V. Ya., Kyrylenko, A.V., Kyseleva, A.H. et al.: Intelligent power systems: elements and modes, pp. 408. Institute of Electrodynamics of the NAS Ukraine, Kyiv (2014)
22. Intelligent power systems: elements and modes: Under the general editorship of acad. of the NAS of Ukraine O.V. Kyrylenko/Institute of Electrodynamics of the NAS of Ukraine, p. 400, (Ukr) (2016)
23. Kuchanskyy, V.V.: The application of controlled switching device for prevention resonance overvoltages in nonsinusoidal modes. In: Proc. 37th IEEE International Conference on Electronics and Nanotechnology (ELNANO 2017), 17–19 April 2017, pp. 394-399. Ukraine, Kiev. https://doi.org/10.1109/elnano.2017.7939785
24. Kundul, S., Ghosh, T., Maitra, K., Acharjee, P., Thakur, S.S.: Optimal location of SVC considering techno-economic and environmental aspect. In: 2018 ICEPE 2nd International Conference on Power, Energy and Environment: Towards Smart Technology 1–2 June 2018 Shillong, India, India, pp. 15-19. https://doi.org/10.1109/epetsg.2018.8658729
25. Zhou, L. Yi, Q., Qin, M., Zhou, L., Zhou, X., Ye, Y.: Using novel unified power flow controller to implement two phases operating in extra-high-voltage transmission system: Proceedings. International Conference on Power System Technology, 13–17 Oct. 2002, pp. 1913-1917. https://doi.org/10.1109/icpst.2002.1067866

26. Kuznetsov, V.G., Tugay, Yu.I.: Trends in the development of power supply systems. Electrical Engineering and Power Engineering, vol. 2, pp. 73–76 (2000)
27. Kuznetsov, V.G., Tugai, Yu.I.: Improving reliability and efficiency of bulk electrical networks. In: Proceedings of the Institute of Electrodynamics of the National Academy of Sciences of Ukraine, K, IED NASU (23), 110–117 (2009)
28. Shidlovsky, A.K., Perkhach, V.S., Skripnik, O.I., Kuznetsov, V.G.: Power Systems with Power Transmission and DC Inserts. K: Naukova Dumka (1992)
29. Gu, S., Dang, J., Tian, M., Zhang, B.: Compensation degree of controllable shunt reactor in EHV/UHV transmission line with series capacitor compensation considered. In: Proceedings of International Conference on Mechatronics, August 29–31, pp. 65–68. Control and Electronic Engineering (MCE 2014), Shenyang, China (2014). https://doi.org/10.2991/mce-14.2014.14
30. Kuznetsov, V.G., Tugay, Yu.I., Shpolianskyi, O.G.: Analysis of internal overvoltage in extra high voltage electrical networks and development of measures for their prevention and restriction Works of the Institute of Electrodynamics of the National Academy of Sciences of Ukraine, vol. 2, pp. 117–123 (2013)
31. Han, B., Ban, L., Xiang, Z., Zhang, Y., Zheng, B.: Analysis on strategies of suppressing secondary arc current in UHV system with controllable shunt reactors. In: Proceedings of IEEE International Conference on Power System Technology (POWERCON), September 28-October 1, 2016, pp. 14–19. Wollongong, NSW, Australia (2016). https://doi.org/10.1109/powercon.2016.7753968
32. Kuchanskyy, V.V.: The prevention measure of resonance overvoltges in extra high voltage transmission lines. In: 2017 IEEE First Ukraine Conference on Electrical and Computer Engineering (UKRCON), pp. 436–441. https://doi.org/10.1109/ukrcon.2017.8100529
33. Joshi, B.S., Mahela, O.P., Ola, S.R.: Reactive power flow control using static VAR compensator to improve voltage stability in transmission system. Proc. Int. Conf. Recent Adv. Innovations Eng. 1–5, Dec. 2016. https://doi.org/10.1109/icraie.2016.7939504
34. Chandrasekhar, R., Chatterjee, D., Bhattarcharya, T.: A Hybrid FACTS Topology for Reactive Power Support in High Voltage Transmission Systems IECON 2018–44th Annual Conference of the IEEE Industrial Electronics Society October 21–23, pp. 65–70. 2018 at the historic Omni Shoreham Hotel, Washington DC, USA (2018). http://doi.org/10.1109/IECON.2018.8591988
35. Tugay, Y.I.: The resonance overvoltages in EHV network. In: IEEE International Conference on Electrical Power Quality and Utilization. Lodz. (1), 14–18 (2009). https://doi.org/10.1109/epqu.2009.5318812
36. Kuchanskyy, V.V.: Application of controlled shunt reactors for suppression abnormal resonance overvoltages in assymetric modes. In: 2019 IEEE 6th International Conference on Energy Smart Systems (ESS), pp. 122–125. https://doi.org/10.1109/ess.2019.8764196
37. Kuznetsov, V.G., Tugay, Yu.I., Kuchansky, V.V.: Overvoltages in single-phase mode. Techn. Electrodyn. (2), 40–41 (Ukr) (2012)
38. Kuznetsov, V.G., Tugay, Yu.I., Kuchanskiy, V.V., Lyhovyd, Yu.G., Melnichuk, V.A.: The resonant overvoltage in non-sinusoidal mode of main electric network. Elec. Eng. Electromech. (2), 69–73, (Ukr) (2018). https://doi.org/10.20998/2074-272X.2018.2.12
39. Hunko, I., Kuchanskyi, V., Nesterko, A., Rubanenko, O.: Modes of electrical systems and grids with renewable energy sources. LAMBERT Academic Publishing, p. 184 (2019). ISBN 978-613-9-88956-3
40. Hunko, I.O., Kuchanskyy, V.V., Nesterko, A.B.: Engineering sciences: development prospects in countries of Europe at the beginning of the third millennium: Collective monograph, vol. 2, p. 492. Izdevniecība "Baltija Publishing", Riga (2018). ISBN 978-9934-571-63-3
41. Zaitsev, I.O., Levytskyi, A.S., Kromplyas, B.A.: Hybrid capacitive sensor for hydro- and turbo generator monitoring system. In: Proceedings of the International conference on modern electrical and energy system (MEES-17) November 15–17, 2017, pp. 288–291. Kremenchuk, Ukraine (2017). https://doi.org/10.1109/mees.2017.8248913
42. Burkov, V.D., Mamedov, A.M., Potapov, V.T., Potapov, T.V., Udalov, M.E.: Fiberoptic sensor of electric field strength. Vestnik MGUL. Lesnoy vestnik. **4**, 130–132 (2008)

43. Kuchanskyy, V., Zaitsev Ie, O.: Corona discharge power losses measurement systems in high- and extra-high voltage transmissions lines. In: Proceedings of the IEEE 2020 IEEE 7th International conference on energy smart systems (2020 IEEE ESS), Ukraine, unpublished (2020)
44. Yakushenko, Yu.T.: Theory and calculation of optoelectronic devices. M, Mechanical Engineering, p. 360 (1989)
45. Okosi, T., Okamoto, K., Otsu, M., Nishihara, X., Kyuma, K., Hatate, K.: Fiber Optic Sensors. Leningrad "Energoatomizdat", p. 256 (1990)

Multifunctional Wireless Automatic Street LED Lighting Monitoring, Control and Management System

Andrii Nazarenko⊙, Zinaida Burova⊙, Oleg Nazarenko⊙, and Anatoliy Burima⊙

Abstract Addressing energy efficiency and saving is one of Ukraine's top energy policy priorities. Reducing energy consumption has a significant impact on reducing imports of fuel resources and increasing the competitiveness of the Ukrainian economy. One of the essential directions is the saving of electricity consumed for lighting the amount of which reaches about 30% the electricity produced total amount. This work describes the Ukrainian scientists research and practical development results aimed at street lighting quality increasing with reducing electricity consumption. The main research objectives to accomplish this goal are formulated. The different types of lighting sources characteristics were analysed and there were demonstrate that modern effective LED light sources have as a number of advantages also negatives. Also, the global problems of LED using to create powerful external lighting sources were declared. The most important problem is the need to provide constant heat removal from the crystal to prevent it from overheating and to increase its service life. The practical developments are the new outdoor lighting sources which has been designed based on high-quality powerful LED modules with minimal power consumption and maximum light output and provides high energy efficiency of lighting networks. Also a multifunctional wireless automatic system for monitoring, control and management of outdoor lighting networks has been developed which not only manages the luminaire or group of luminaires on/off depending on the ambient light level, but similarly controls the temperature of the LED crystal and the lamps power supply to prevent them from overheating, which provides the luminaires longer service life.

A. Nazarenko (✉) · O. Nazarenko
Monitoring and Optimization of Thermal Processes Department, Institute
Of Engineering Thermophysics NAS of Ukraine, 2a, Marii Kapnist Street, 03057 Kyiv, Ukraine
e-mail: nazarenkoandrii21@gmail.com

Z. Burova
National University of Life and Environmental Sciences of Ukraine, 15, Heroyiv Oborony Street, 03041 Kyiv, Ukraine

A. Burima
Delta SOFT LLC, Kyiv, Ukraine

V. Babak et al. (eds.), *Systems, Decision and Control in Energy I*, Studies in Systems, Decision and Control 298, https://doi.org/10.1007/978-3-030-48583-2_9

Keywords Energy saving · Streetlight network · LED module · Automatic control and monitoring system

1 Introduction

Improving quality and living standards is the goal of Ukraine's national regional policy. It's first priority is "…creation and maintenance of a complete living environment, improving the quality of peoples life" [1]. One of the main ways to do this is to create efficient and reliable outdoor lighting networks, to ensure the energy saving, proper functioning and storage of lighting fixtures on site. At the same time, street lighting costs amount to about one third of the total amount of electricity consumed [2]. Therefore, for a quality realization of the state regional development policy it is equally important to provide energy conservation and the implementation of energy efficient technologies and modern lighting equipment.

Modernization of street lighting networks of Ukrainian cities and towns has a significant role in the law thesis implementation, which will certainly increase the comfort of life, reduce the population traumatization due to weather conditions and poor visibility on roads and prevent emergencies and criminogenic situations, especially in the dark. According to experts, the street lighting systems in Kyiv do not meet current requirements and state standards. The city needed to install about 5–6 thousand additional points of street lighting, improve the management and control of the outdoor lighting networks by introducing equipment for remote control and prompt correction the outdoor lighting system parameters. For small settlements, their street lighting is often obsolete, inadequate or absent. Also, the street lighting networks modernization requires investment.

Solving this problem requires an integrated approach. Reducing the consumption of electricity used to illuminate streets, roads and surrounding areas is possible through the use of high-quality modern luminaires with minimal energy consumption and optimum optical characteristics. For the street lighting networks proper functioning it is necessary to provide an independent monitoring of their status and to implement an automatic control system of the parameters a separate light source and the whole network as one. Such complex of theoretical researches has been realized by specialists Monitoring and Optimization of Thermal Processes Department Institute of Engineering Thermophysics NAS of Ukraine where already are practically developed and implemented:

- A modern street lamps based on powerful LED modules,
- Multifunctional wireless automatic street lighting monitoring, control and management system.

2   Main Part

2.1   Formulation of the Problem

To accomplish the goal of reducing electricity consumption used for illumination of street roads and surrounding territories, due to using the high-quality lamps with minimal energy consumption and maximum light flux, implementation of street lighting monitoring and control systems the following main research objectives were formulated:

1. to carry out the existing lighting sources comparative analysis and to determine the main parameters that affect the luminous flux and lifetime;
2. to develop a lighting device using high quality light sources, which will allow to provide maximum light flux with minimal electricity consumption;
3. to create a wireless automatic control and monitoring system for street lighting based on the developed device;
4. to solve problems of the roads and adjoining territory illumination in places with no centralized electricity supply network;
5. consider ways to solve the problem of heat dissipation from LEDs.

2.2   The Lighting Sources Characteristics Review

The way to reduce electricity consumption used to street illumination is to use high-efficiency lamps that are designed to maximize light flux and lifetime while minimizing energy consumption. Consider the characteristics of the most common lamps types used for street lighting. According to professional studies [3–5], the electricity amount that is converted to light in a vacuum incandescent lamp is only 7–10%, the rest of the power is transformed into thermal infrared and other types of radiation. Fluorescent lamps generate 21% of visible light. Halogens create a ray in the visible range of which is only 27% of the total radiation intensity. LEDs only generate visible rays. The entire light temperature range of LED lamps is in the range of 3000...6500 K or 400...700 nm—from red to blue. We can compare the characteristics of light sources different types from the data presented in Table 1.

We can admit that incandescent lamps are no longer used in modern street lighting due to their low efficiency. Currently, discharge lamps working by passing electricity through mercury vapour, which in turn emits ultraviolet light, are widely used in street lighting. Gas-discharge lamps replaced the incandescent lamps, however, their use is threatened by a number of disadvantages, such as monochromatic yellow light (high or low pressure sodium lamps), flickering of light leading to eye fatigue, duration of operation depends on the number of on/off cycles, ballast noise, mercury vapour damage due to possible rupture, the difficulties of re-ignition for high-pressure lamps. Despite the drawbacks the discharge lamps were preferable to incandescent lamps in many lighting applications prior to improvements in LED lamp technology [5]. The

Table 1 The lighting sources characteristics

The light element type	Incandescence	Luminescent	Gas-discharge	LED
The luminous flux efficiency, Lm/W	10–15	30–45	50–90	90–210
Working life, h	5000	6000–10000	6000–10000	30000–100000
Stroboscopic effect (ripple), %	18%	20–30%	20–30	<5%
Power factor PFC, cos φ	0,95	0,6–0,8	0,6–0,8	0,95

Table 1 data prove that the LED lamps are the most effective in creating a powerful and bright luminous flux, do not flicker and have the longest working life which is what is needed in the creation of outdoor lighting sources. The LED luminaire price is higher than similar discharge lighting devices, but the benefits of using them quickly pay off in the short term.

2.3 The Main Problems of LED Using

So, LED light sources have a number of undeniable advantages, such as high light output, low control voltages, long product life, mechanical durability and reliability, high electrical safety and environmental friendliness. At the same time, the disadvantages of the current LED devices include severe temperature limitation of light-emitting semiconductor crystals (from 85 to 125 °C) [6, 7], which exceeds the sharp degradation of the LED's optical parameters.

According to the researches results, the following global problems of LED using to create powerful external lighting sources can be formulated [8]:

1. The LEDs energy efficiency increasing problem, which requires the creating heteroepitaxial structures technology improving, the quantum yield and efficiency of phosphors increasing, the thermal resistance optimization. Leading developers of LED modules are working on its solution.
2. The problem in the physical principles of thermoregulation of the crystal through which sufficiently large current passes, heating it, which leads to the LED's degradation. Therefore, it is necessary to effectively remove heat by creating the radiator systems. It is economically feasible to replace all the metal parts of the heat extraction system with heat-conducting ceramic or plastic ones. This requires the development and creation of new systems for light-emitting crystals thermal stabilization based on new heat-conducting materials.
3. The following problem is related to the electronic control principles of LED light sources and requires the development of special devices (drivers) with the appropriate efficiency, high power factor, reduced noise and optimized sizes.
4. The last problem is related to the optical principles of the light fluxes' direction formation. Solving this problem will allow to implement different types of light

distribution LED lighting systems and provide them with modern ergonomic parameters.

The problem of heat dissipation is one of the main in creating powerful luminaires for outdoor use based on LED. For stable and long-lasting operation, they require constant removal of heat energy from the crystal—the radiator. Constant overheating of the light-emitting crystals at times reduces the semiconductor device's life, contributes to a brightness smooth loss with the working wavelength displacement.

Structurally, all radiators can be divided into three groups: plate, rod and ribbed. In all cases, the area near the LED to the radiator may be in the form of a circle, square or rectangle. The thickness and area of the radiator is of fundamental importance when choosing, as these characteristics are responsible for the removal and uniform distribution of heat throughout the surface of the radiator.

Currently, the cooling of powerful LEDs is produced mainly on aluminium radiators. This choice is due to the lightness, low cost, machinability and good heat-conducting properties of this metal. Its thermal conductivity coefficient is in the range of 202...236 W/(m K) and depends on the purity of the alloy. By this characteristic it is in 2,5 times higher than iron and brass. In addition, aluminium is subjected to various types of machining. To increase heat dissipation properties, an aluminium radiator is anodized.

In household LED lighting devices, the radiator is usually the body of an LED lamp with which electronic control circuit of the LED device (driver) inside. Institute of Engineering Thermophysics in collaboration with the Institute of Semiconductor Physics NAS of Ukraine work was carried out on mathematical modelling of thermal mode of cooling systems of LED lamps based on thermally conductive ceramics and plastics. The analysis of the temperature distribution in the elements of the lamp showed that lamp's heat sink system provides the required thermal regime of the LED module. Numerical solution confirmed the efficacy of LED lamp's body-core to divert heat from the LED module [7]. The use of new composite materials (metal ceramics) in high-power industrial and outdoor LED luminaires is currently under development to further study their characteristics and predict their change during long-term operation.

3   Results

3.1   Street LED Lighting Source Designing

Modern technologies make it possible to produce LEDs of different sizes and capacities. There are many types of configurations of their location in lighting devices. It depends on the consumers' needs and the further operation conditions. It is clear that for the street lighting needs it is necessary to choose powerful LEDs that provide maximum light flux, have higher than the ordinary household lamps maximum permissible operating temperature at which there is no crystal degradation, long life expectancy and reasonable price. Leaders in the LEDs production consider the United States, Germany and Asia countries [5].

It should also be noted that lamps for streets, parks and roads must meet many criteria. The main features to consider when designing new luminaires are the following:

1. Power saving. Street lights illuminate large areas, and it is especially important that most of the light emitted by LED devices is directed to the illuminated surface. They allow energy savings compared to similar high-pressure discharge lamps and sodium lamps.
2. Structural strength and environmental protection. The case of the device should be designed so that debris, water and other pollutants do not accumulate on the surface of the lamp and do not impair its cooling ability, transparency of the protective glass, thereby preserving characteristics throughout the entire service life.
3. Colour rendition. LED lighting sources have the best colour rendering characteristics, the colour tone and rendering index can be selected for a specific lamp application.
4. The durability of LED lamps is significantly longer than the traditional street light sources, but LED light sources are sensitive to raised temperatures. The manufacture of a metal lamp case, which simultaneously serves as a radiator and provides good heat dissipation, from a highly heat-conducting material will extend the life of the lightbox.

To develop a new powerful source of external lighting based on LED luminaire, a thorough study the characteristics the nomenclature of LEDs presented in Ukraine was analysed (Table 2). The main selection parameters at the optimal price/quality ratio were the light flux, the working life, the maximum allowable operating temperature at which the crystal does not break.

Table 2 LED characteristics

Manufacturer	Cree, LG, Osram, Seoul Semiconductor [9]	Epistar, Prolight, SemiLeds and other Chinese origin brands
Luminous flux, Lm/W	up to 210	80–90
Working resource, h	up to 100 000	10 000–30 000
The substrate temperature at the rated current applied to the LED, °C	50–65	80–105
The maximum permissible temperature of operation at which no crystal degradation occurs, °C	120–130	100–115
Static breakdown protection	present	not existing
Manufacturer's warranty, years	3–5	1

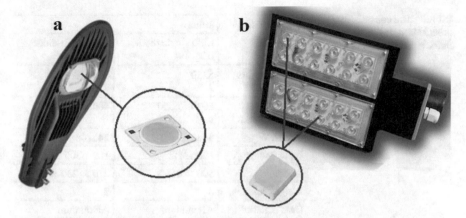

Fig. 1 Console LED single diode **a** and multi-diode **b** street luminaries

As we can see, Seoul Semiconductor Korean LED modules [9] have been selected because their use allows to increase the luminous flux almost twice in comparison with similar lamps created on the basis of components from China.

Based on these LEDs, two luminaire designs have been developed [8, 10]: single and multi-diode, presented in Fig. 1. In Fig. 1a shows a 35 W single diode lamp. The LED module with the driver, power supply, sensors and control system are installed in the standard EFA case which is made of high-conductive aluminium alloy and has a developed radiator system to provide efficient heat removal from the crystal and electronic components of the lamp.

The lamp in Fig. 1b is made on the basis of two boards with 12 LEDs 2 W capacity placed on each. Its case is made of high-quality aluminium alloy and polycarbonate diffuser. The advantage of this model, based on a multi-diode array, is that in the event of a single diode failure the lamp continues to operate, and due to more diodes, it has a better heat dissipation surface.

The main characteristics of the developed luminaries are presented in Table 3.

3.2 Wireless Monitoring, Control and Management System Description

Light networks created on the powerful LED-basis luminaries meet all the requirements the modern approach to illuminate industrial areas, streets, roads, surround territories, etc. LED lamps are already energy efficient light sources, but even more energy saving can be achieved by adjusting their power mode depending on the time of year or time of day. Also, as mentioned above, it is necessary to tightly control the temperature regime of the LED in real time, since the light-emitting crystals constant

Table 3 The console LED street luminaries' main characteristics

Characteristic, measure	Value	
	LED single diode (Fig. 1a)	LED multi-diode (Fig. 1b)
Luminous flux, Lm	5250	3600…7200
Colour temperature, K	4000…6500	4000…6500
Power, W	35	24…48
Input voltage, V	198…242	100…305
Size, mm	500 × 215 × 75	250 × 230 × 50
Weight, kg	3	2
Case material	Aluminium	Aluminium
Protection rating	IP65	IP65

overheating at times reduces the life of the semiconductor device, resulting in the brightness loss with the working wavelength shifting.

A multifunctional Monitoring, Control and Management System has been developed to provide the street lighting networks operative administration. It gives the all necessary information about the temperature parameters of luminaires and the environment and allows to adjust the LED lamps brightness either manually or automatically and individually or for group depending on the ambient lighting and pre-set work algorithm.

The system schematic block diagram is presented in Fig. 2.

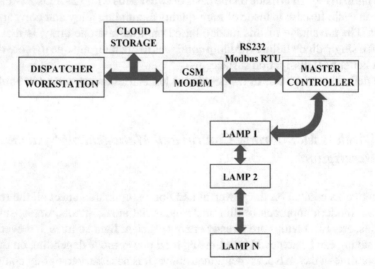

Fig. 2 Multifunctional street LED lighting Monitoring, Control and Management System block diagram

The control system consists of the Dispatcher Workstation, Master Controller and N-number of LED-lamps. Dispatcher Workstation consists of a PC with an installed management and monitoring program.

The Master Controller is a hardware-software complex based on a high-performance microcontroller with an integrated radio module. Each lamp equipped with a control module consists of a high-performance microcontroller with an integrated radio module, temperature and light sensors, a power supply system and a lamp driver control.

The wireless radio network is built on the principle of the MESH network while maintaining communication at the stack level of the radio modules. After switching on the device sends a special message to the air and waits for a response from devices from its subnet and establishes a connection with them. If there are several devices in the detection zone from its own subnet, the connection is established with the one with whom the test message transmission time is less. If there are no devices in the network coverage area to establish communication, the device falls asleep for 1 min and then sends a test message again. If during operation the connection with the device with which the connection was established is lost, reconnection occurs.

The following connection types can be used as a transmission channel:

- wired (RS 232, MODBUS RTU)—with direct connection of the Dispatcher Workstation with the Master Controller;
- converter TCP < > COM—(TCP/IP, MODBUS RTU over TCP)—with a remote connection of the Dispatcher Workstation with the Master Controller;
- GSM modem—(TCP/IP, MODBUS RTU over TCP)—with a remote connection of the Dispatcher Workstation with the Master Controller.

The control commands via the MODBUS protocol through the selected transmission channel are sent to the Master Controller, which converts the received command into a telegram and sends it to the radio network. Since telegrams for various purposes (configuration, control, group, individual) are transmitted to the radio network, the structure of the telegram can have a different format, for example, group telegrams are transmitted by broadcast message, but are processed only by devices that belong to the group of which the message is intended. A message intended for a particular device is transmitted addressable with the receipt of a confirmation message. The passage of a group message is monitored by a confirmation message from the most recent (territorial) device in the group designated as the "group monitor".

Addressing devices in the system occurs on the basis of a unique identifier specified during the production of the microcontroller. If the device cannot execute the received command, it sends a message indicating the malfunction. When transmitting each message, accurate time signals are transmitted, so when transmitting any message, an additional time adjustment in the luminaires takes place to ensure synchronized operation in the scheduled mode.

The Master Controller provides the ability to remotely reboot the router, monitor lighting and temperature at the installation site, and control an additional relay output. Non-volatile real-time clocks are provided with the possibility of adjustment from the Dispatcher Workstation.

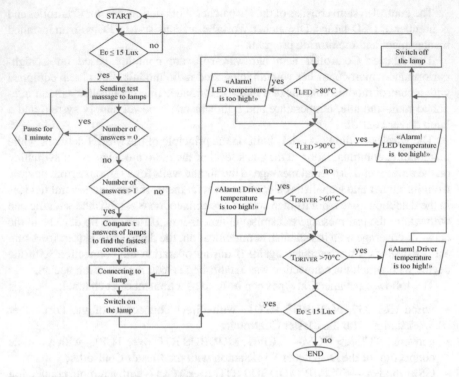

Fig. 3 LED-lamp network operation algorithm

After two-way radio communication is installed, the operation of each lamp in the system is managed by the algorithm shown in Fig. 3.

As noted above, each luminaire is equipped with a light sensor. When the outside lighting falls below 15 lx, the lamp switches on at 10% power. With ambient light level falls every 1 lx lower the luminaire increases its power by 10% and so to 100% output. The illumination quantity is monitored every 30 s. Accordingly, when the ambient light level grows up above 15 lx the lamp is automatically switched off.

Also, the temperature is constantly monitored by sensors located directly on the LED and on its electronic control scheme—driver. When the LED's temperature sensor registers a value that reaches the warning set limit—if the lamps crystal temperature rises more than 80 °C, the device sends a warning message to the Dispatcher Workstation, when crystal temperature reaching the emergency setting (more than 90 °C), the lamp is switched off and an alarm message is also sent to the operator. The driver's temperature is the same controlled. If its temperature is higher than 60 °C, the dispatcher will receive a warning message, and if more than 70 °C, the lamp will switch off and the operator will receive an alarm re-notification.

All pre-sets (initial ambient light level, critical crystal and driver temperature values, polling frequency) can be changed remotely. Critical temperature values are usually programmatically set 25...30% below the passport value.

Fig. 4 Monitoring, Control
and Management System
main window

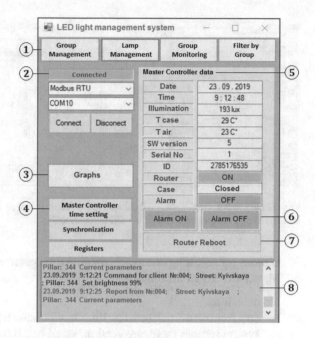

3.3 Monitoring, Control and Management System Software

All monitoring data is stored in the cloud storage and displayed on the Dispatcher
Workstation PC screen. Dispatcher Workstation software is developed in C#. Development environment is Microsoft Visual Studio. To implement the Modbus RTU
protocol free third-party libraries were used. Storage of configuration data, as well
as all the information necessary to provide the operation of the system, is stored in
the SQLite database.

The developed system's main window is presented in Fig. 4.

The main program window contains groups of items according to marks in Fig. 4:

1. The command buttons for activating program windows: Group Management,
   Lamp Management, Group Monitoring, Filter by Group. Each of these windows
   contains many general and special programmatic settings for the corresponding
   parameters, indication of feedback from devices and their groups, and makes it
   possible to remotely and operatively control every unit.
2. Modem connection status:

   a. drop-down menus for selecting a connection protocol (MODBUS RTU) and
      com port number;
   b. command buttons for establishing and ending communication
      (Connect/Disconnect).

3. The button for the chart window of environmental parameters, light level and
   temperature, real-time monitoring.

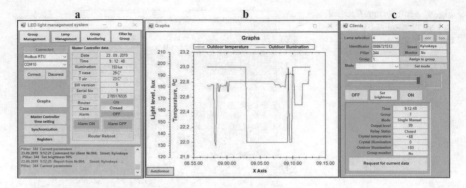

Fig. 5 Dispatcher Workstation data display: **a**—Main window; **b**—environment Graphs: Light level (red) and Temperature (blue); **c**—Clients window

4. Buttons for adjusting and synchronizing time parameters.
5. Master Controller data interface:

   a. date, time;
   b. data from measuring sensors: illumination, case temperature, air temperature;
   c. system settings (software version, serial No, ID);
   d. the router, case and alarm status indicators.

6. Alarm On/Off buttons.
7. Remote reboot router button.
8. Real-time Events Window.

A typical display of software information windows on the Dispatcher Workstation PC screen is shown in Fig. 5.

In addition to the Main window (a), where the parameters of a certain luminaire are already shown in the Master Controller data interface, it usually includes a window for monitoring environmental parameters—Graphs (b) for that light source, as well as a Clients window (c) with variants of configuration in its extended functions showed in Fig. 6: 1—single, 2—group, 3—input to Database, 4—Group monitor On/Off.

Developed system can implement an automatic or manual control a separate lamp or luminaire groups (for example, streets, street side, through one for economical use). It also performs the following functions:

- Control depending on the ambient lighting, season or day time.
- Monitoring the temperature of the LED-module (bough matrix and driver) which will prevent the timely failure of the lamp.
- Monitoring the light flux of the matrix which will allow timely automatic detection of a decrease in light flux or failure of the lamp.
- Completely disconnecting the lamp from the power supply during the non-working period.

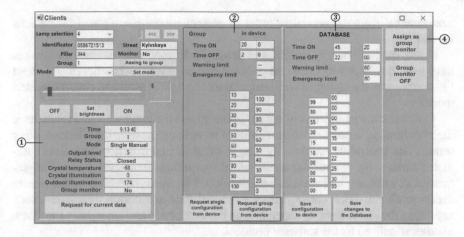

Fig. 6 Clients window with variants of configuration

At the same time, the implementation of the developed monitoring, control and management system of the street lighting networks does not require the laying of additional communication lines, make available remote management of the system via the Internet, provides control and settings from a PC with the ability to create their own GUIs.

At the moment, the software is designed to work with one main controller, but if necessary, it can be expanded to the required number of serviced main controllers, which will allow you to manage an almost unlimited (within reasonable limits) number of devices. It is possible to expand the functions of management and data display.

It is also additionally possible to expand the functions of terminal devices for the implementation of specific tasks in parallel with lighting control, for example, a system for monitoring gas pollution by exhaust gases, monitoring the radiation environment, temperature and humidity of the environment and other parameters.

4 Conclusions

A modern approach to the street lighting networks creation has been implemented as a result of complex research and practical developments of Ukrainian engineers aimed at reducing electricity consumption for lighting outdoor areas. An analysis of the light sources used to create modern outdoor street lighting systems has shown the increasing superiority of LED lamps. Their indisputable advantages are the white light flux high efficiency with minimal energy consumption, a significant service life, and the flickering absence. Along with the advantages, LED light sources have several negative factors, the predominance of which is the need to ensure constant heat removal from the crystal in order to prevent its overheating and increase its service life.

Based on high-quality single and multi-diode modules, modern luminaires have been developed to create an outdoor lighting system. The developed street lighting sources are equipped with temperature and light sensors and communication modules.

To monitor the parameters of the environment and the luminaires and for manage their functioning depending on the results of processing the signals of the measuring sensors and control the outdoor lighting network, a multifunctional wireless auto-mated monitoring, control and management system has been developed and imple-mented. A software shell also has been developed for this system. It provides remote access to all networks' components for the real-time monitoring their parameters, regulating the on/off lighting depending on the time of day and emergency shutdown in cases stipulated by the software protocol.

References

1. LAW OF UKRAINE The Principles of State Regional Policy, https://zakon.rada.gov.ua/laws/show/156-19, last accessed: 2019/12/22
2. Ożadowicz, A., Grela, J.: Energy saving in the street lighting control system—a new approach based on the EN-15232 standard. Energy Eff. **10**, 563–576 (2017). https://doi.org/10.1007/s12053-016-9476-1
3. THE PROMISE AND CHALLENGE OF LED LIGHTING: A PRACTICAL GUIDE: LED Handout, Rev. 6/21/18. https://www.darksky.org/wp-content/uploads/bsk-pdf-manager/IDA_LED_handout_48.pdf, last accessed: 2019/11/12
4. Learn About LED Lighting, https://www.energystar.gov/products/lighting_fans/%20light_bulbs/learn_about_led_bulbs, last accessed: 2019/11/06
5. Small lights with big potential: light emitting diodes & organic light emitting diodes. Commercial History (1960s—Today), https://edisontechcenter.org/LED.html, last accessed: 2019/12/02
6. Lothar, N.: Cooling and temperature control of LEDs. Solid-State Lighting, 2010. No. 3. https://www.led-e.ru/articles/led-cooling/2010_3_13.php, last accessed: 2020/01/15
7. Basok, B.I., Davydenko, B.I., Sorokin, V.M., et al.: Numerical modelling of the thermal regime of LED lamps. Ind. Heat Eng. **36**(5), 10–23 (2014)
8. Burova, Z., Nazarenko, A., Nazarenko, O.: The main problems of developing powerful led systems for outdoor use,. MEIT **09–01**, 118–122, Oct. (2019). http://di.org/10.30890/2567-5273.2019-09-01-028
9. Seoul Semiconductor. Products, http://www.seoulsemicon.com/en/product/, last accessed: 2020/02/11
10. Burova, Z., Nazarenko, A., Nazarenko, O.: Street Lighting Networks Modernization Using Energy Efficient LED Sources, SWorldJournal (2), part 2, 45–49 (2019). https://www.sworld.com.ua/swj/swj02-02.pdf, last accessed: 2020/02/11

Emergencies at Potentially Dangerous Objects Causing Atmosphere Pollution: Peculiarities of Chemically Hazardous Substances Migration

Oleksandr Popov⊕, Dmytro Taraduda⊕, Vitalii Sobyna⊕, Dmytro Sokolov⊕, Maksym Dement⊕, and Alina Pomaza-Ponomarenko⊕

Abstract The most chemically dangerous technogenic objects on the territory of Ukraine are thermal power plants, combined heat and power plants, nuclear power plants, enterprises of chemical industry and metallurgy. Various pollutants are released into the atmosphere through chimneys of different heights due to their operation. Emergencies can occur due to significant pollution of the surface of the atmosphere in adjacent areas due to technological disturbances or adverse weather conditions, or unauthorized emissions, etc. Preventive measures are based on the application of mathematical models and related software for the propagation of harmful impurities in the atmosphere. Structure of new information and technical methods of such emergencies prevention is presented. Given work describes the most likely occurrence and development of emergencies related to entry of toxic substances into the atmosphere at these potentially dangerous objects. Diagram shows the scheme of migration of impurities in the air due to technogenic emissions. Influence of the main factors on scattering of impurities under continuous and volatile emission conditions is described. It is shown that the greatest influence on distribution of concentration of dangerous substances in the atmospheric air is caused by: source characteristics (source type, mode and conditions of emission), meteorological characteristics (wind direction and velocity, atmospheric stratification, precipitation, temperature and humidity), pollutant characteristics (ability to interact chemically with other substances in atmospheric air, gravitational deposition rate, absorption coefficient of the underlying surface), characteristics of the earth's surface of adjacent area (topography, roughness). Obtained results are basis for the further development of new mathematical models of atmospheric pollution from emissions of chemically hazardous objects. It is necessary for effective solution of the problems to prevent such emergencies.

O. Popov (✉)
Pukhov Institute for Modelling in Energy Engineering of NAS of Ukraine, Kyiv, Ukraine
e-mail: sasha.popov1982@gmail.com

State Institution "Institute of Environmental Geochemistry" of NAS of Ukraine, Kyiv, Ukraine

D. Taraduda · V. Sobyna · D. Sokolov · M. Dement · A. Pomaza-Ponomarenko
National University of Civil Defence of Ukraine, Kharkiv, Ukraine

V. Babak et al. (eds.), *Systems, Decision and Control in Energy I*, Studies in Systems,
Decision and Control 298, https://doi.org/10.1007/978-3-030-48583-2_10

Keywords Potentially dangerous object · Emergency · Air pollution · Factors of influence

1   Introduction

Anthropogenic environmental pressure in Ukraine led to significant increase in risk of technogenic and natural emergencies. They should be understood as a violation of normal living conditions and activities of people in particular territory or object on it or in a water body. It can be caused by an accident, catastrophe, natural disaster or other dangerous event, including an epidemic, an epizootic, an epiphytosis, a fire that can lead to inability of population to reside in the territory or object and conduct there business activity. It can be also associated with loss of life and/or considerable material damage.

Every year, there are large number of emergencies in the country. They cause great material damage and lead to human victims. The greatest danger today in Ukraine is accidents: radiation, chemical emissions of chemically and biologically dangerous substances, hydrodynamic, transport, energy systems and treatment facilities; fires, explosions, possible earthquakes and all kinds of dangerous geological manifestations.

Progress and results of the work on elimination of emergencies depend not only on the ratio of resource-economic, moral-political, scientific-technical and organizational potentials of the system, but also on efficiency of the operational management. Efficiency of decision-making process, implementation and correction is determined by ability of management to focus its main efforts on the main directions to eliminate causes of emergencies and its consequences.

Management basis is decision of emergency liquidator who is fully responsible for managing of subordinated forces and successfully fulfilling their tasks to eliminate emergency consequences. Management consists in resolute and persistent implementation of measures envisaged by the Emergency Response Plan and taking decisions on implementation of the Emergency Response tasks within established timeframe.

Efficiency of the management system to prevent emergencies of natural or technogenic nature is achieved by the use of modern mathematical approaches and methods of decision-making supporting.

2   Literature Analysis and Problem Statement

Aspects of this problem were considered and solved by many well-known scientists. Borsdorff T., Carn S., Dickerson R., Yang K., Mlakar P., Lysychenko G., Kovalenko G., Zabulonov Y., Verma S., Wei J., Tsiouri V., Kovalets I., Ievdin I., Gong P., Tang X., Huang X., Pereira M., Schirru R., Gomes K. and others were involved in addressing of environmental safety of potentially dangerous objects (PDOs).

Issues of economic evaluation of prevention, elimination and consequences of emergency, development and use of telecommunication facilities and methods and unmanned aerial vehicles to eliminate emergencies are discussed in [1–8]. Features of construction of emergencies monitoring systems of different complexity are described in [9–13]. There are number of means to prevent emergencies of techno-genic nature and they are highlighted in works [14–17]. Publications [18–26] are devoted to development of conceptual approaches and technical means to solve prob-lems of emergency prevention on the territories o of potentially dangerous objects. However, little attention was paid to development of efficient methods to prevent natural and technogenic emergencies related to chemical pollution of atmospheric air (AA) in the areas of PDO deployment in the field of civil protection. Therefore, the work on developing of new prevention methods of such emergencies is relevant, timely and important for Ukraine.

The authors of this publication began the work on the development of new infor-mation and technical methods of preventing emergencies of natural and technogenic nature associated with chemical contamination of AA in the territories of PDO taking into account given above information.

The information-technical method is a method that allows us to solve the given task by completing of following five steps:

1. analyzing physical (or chemical, biological, etc.) features of functioning of researched object;
2. development of mathematical model of the researched object;
3. development of information-computing procedures that allow us to realize developed mathematical model;
4. development of control algorithm that implements appropriate procedures;
5. development of hardware and software for practical implementation of proce-dures according to appropriate control algorithm.

3 Purpose and Objectives of the Study

The article aim is to research physicochemical characteristics of distribution of toxic substances in the atmospheric air during emergency at potentially dangerous object. The following tasks were set and solved to achieve this aim:

1. Description of the most likely emergencies scenarios related to release of toxic substances into the atmosphere at potentially dangerous object.
2. Development of conceptual scheme for distribution of pollutants in the surface atmosphere. Analysis of the main influence factors on the process under stationary and non-stationary emission conditions.

4 Research Results

The most chemically dangerous technogenic objects on the territory of Ukraine are thermal power plants, combined heat and power plants, nuclear power plants, enterprises of chemical industry and metallurgy. Their operation leads to formation of various pollutants. They are released into the atmosphere through chimneys of different heights. Such emission sources are stationary point objects according to the principles of mathematical modeling [27].

Emergencies associated with significant atmospheric air pollution on the territories of the above-mentioned PDO can occur both in regular and emergency mode of their operation. In the first case emergency will arise under the following conditions:

(1) PDO emits (fast, short-term or continuous) impurities into the atmosphere according to the regulation, but meteorological conditions are such that the pollutants do not have time to dissipate. They rapidly transferred by the wind to the earth's surface, where their concentrations is become much higher than corresponding limit. In this case emergency should be responded quickly and effectively by the civil protection units to the relevant PDO.

(2) PDO carries out unauthorized emissions that exceed regulatory standards. In this case even under favorable meteorological conditions level of surface concentration of toxic substances will exceed MPC. It will cause emergency.

Under emergency mode of operation the PDOs usually carry out powerful unregulated emissions into the atmosphere. It means that the levels of pollutants concentrations near the earth's surface at a certain distance from the emission source will significantly exceed corresponding MPCs. This situation is also an emergency that needs to be addressed in timely and effective manner.

Cases of emergencies caused by significant chemical contamination of atmospheric air are one of the most common and dangerous. Efficient prevention of such emergencies requires detailed study of physical characteristics of distribution of dangerous substances in the atmospheric air. It will be discussed below.

4.1 Factors Influencing Distribution of Impurities in the Atmosphere

Continuous emission. Nature of impurities movement in the air is determined by their own physical and atmospheric properties. The conceptual behaviour of pollutants emitted by stationary point technogenic source is shown in Fig. 1 [28].

The main factors influencing distributions of impurities in the atmosphere are wind and distribution of air temperature by altitude.

Industrial emissions into the atmosphere have certain rate of escape from pipes. They are characterized by buoyancy in the case of gases overheating in relation to ambient air. Therefore, vertical velocity is created in the vicinity of discharge

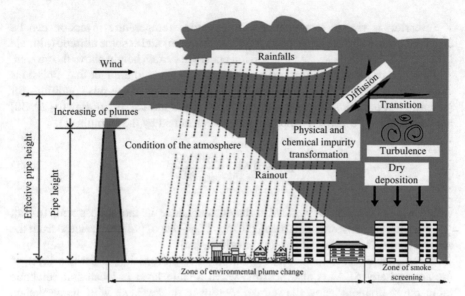

Fig. 1 Peculiarities of pollutants distribution in the atmosphere caused by emission of technogenic source

source. It decreases with distancing from the source, sometimes extending over long distances and contributing to impurity uplift.

Process of impurities transferring to the upper atmosphere is determined by the category of atmosphere stability (stratification of the atmosphere).

The atmosphere state can be equilibrium, stable and unstable. Degree of stability of the atmosphere determines behavior of air particle (elemental volume of air).

Part of warm air rises up, cold part—goes down. Usually in the atmosphere there is a decrease in temperature with altitude. Air at any altitude is in equilibrium if temperature gradient in dry atmosphere is 1 °C per 100 m. Each upward or downward air particle receives ambient air temperature and its density becomes equal to density of the surrounding particles. So there is no reason to raise or lower it. In this case, it is said that the atmosphere is in state of indifferent equilibrium or that there is equilibrium stratification. Usually, in humid conditions, equilibrium is observed at lower temperature gradient (approximately 0.6 °C/100 m).

Upward moving particle will cool and soon become colder than surrounding air and heavier than it if vertical temperature gradient in the atmosphere is less than at equilibrium. Therefore, it will fall and again take its original position. In this case, the atmosphere is in state of stable equilibrium, i.e. there is a stable stratification of the atmosphere.

Air particle will continue to move with increasing acceleration as it starts to move up or down if vertical temperature gradient is more than equilibrium. Then it goes from its original position. In this case, it is called unstable stratification [29].

Inversion is rise in temperature with altitude. Temperature inversion can be observed both at surface of the earth (surface inversion) and at some altitude (altitude inversion). It is called elevated if inversion occurs at certain height above the ground.

Presence of increased inversion over the source leads to the fact that emissions cannot rise above a certain level—the "ceiling". It creates dangerous conditions for increasing of pollution concentration near the earth's surface. According to [20, 30] the height of the "ceiling" is approximately determined by the formula (1):

$$z_c = 0.61 \sqrt{\frac{W_0 R_0^2 \Delta T}{K_z \frac{\Delta T}{\Delta z}}} \tag{1}$$

where K_z—coefficient of vertical turbulent exchange in the atmosphere, $[m^2/s]$; $\frac{\Delta T}{\Delta z}$—inversion temperature gradient, $[°C/m]$; W_0—rate of pollutant release from the source, $[m/s]$; R_0—radius of the source orifice, $[m]$.

Significant increase in the concentration of impurities in the surface layer of the atmosphere (SLA) is also possible when a calm layer is located at adiabatic temperature gradient below the source. Moreover, thicker layer with the weakened wind speed and with lower location leads to stronger impact.

However, released gas mixture into the atmosphere has certain velocity and certain buoyancy due to temperature difference of mixture and the ambient air. So, discharged emission continues to move for some time in the direction specified by the pipe. Process of diffusion scattering will begin only after stop of such directional movement. In other words, under real emission conditions there is fictitious source of diffuse impurities in the atmosphere, which is raised above orifice at some height.

Concept of efficient height of the source is appeared:

$$H_{ef} = H_s + \Delta H \tag{2}$$

where H_s—emission source height (flare pipe), (m); ΔH—magnitude of impurity rise above the source, (m).

Formula for calculation ΔH for different states of the atmosphere can be found in [31].

Dispersion of impurities in the atmosphere is observed due to scattering casued by turbulent diffusion and wind transfer.

Impact of wind causes formation of ejection torch to form during continuous output from the impurity source. Diffusion transfer may prevail over the wind in light winds or in the absence of wind. In such case calm cloud of pollution is formed around the source.

Effect of wind speed on the pollution of SLAs is complex because ground concentrations of impurities decrease with increasing wind speed and increased wind leads to decrease in the initial rise of impurities which contributes to increase in ground concentration. There is a certain dangerous wind speed at which maximum concentration of impurities is observed in the case of impurities overheating relative to the outside air. This value u_m (m/s) at the level of the weather vane for source of comparatively heated emissions is approximately determined by the formulas [29]:

$$u_m = \begin{cases} 0.5 & at \quad v_m \leq 0.5m/s, \\ v_m & at \quad 0.5 < v_m \leq 2m/s, \\ v_m\left(1 + 0.12\sqrt{f}\right) & at \quad v_m > 2m/s, \end{cases} \tag{3}$$

where $v_m = 0.65\sqrt[3]{\frac{V\Delta T}{H_s}}$, $V = \pi R_0^2 W_0$, $f = 2000 \cdot \frac{W_0^2 R_0}{H_s^2 \Delta T}$.

Increased levels of pollution will be observed in low winds (0–1 m/s) due to the accumulation of impurities in the SLA with impurities colder than outside. Therefore repeat of low wind speeds and repeat of dangerous wind speeds should be taken into account during research of dispersion conditions.

Irregular air movements (atmospheric turbulence) are created in SLA at vertical temperature gradient of much higher than 1 °C/100 m.

Turbulent diffusion of impurities in the air is observed due to the influence of turbulent vortices on emission cloud. Nature of their interaction significantly depends on the relative size of the vortices and the cloud. Figure 2 shows idealized scatterplots that qualitatively illustrate this process. The dashed line shows the trajectory of the cloud center [32].

Circular cloud is shown in Fig. 2a. It is in the field of turbulent vortices whose dimensions are smaller than its diameter. Cloud in this case increases and concentration of impurities in it decreases. Figure 2b shows the interaction of small cloud with large-scale eddies. In this case result of the turbulence effect is to distort the cloud trajectory without changing concentration of impurities in it. Cloud in a vortex field is shown Fig. 2c. Its dimensions are correlated with the cloud diameter. In this case diffusion is intense. Concentration of impurity in the cloud decreases rapidly and its shape is noticeably and randomly distorted often breaking up into separate clubs.

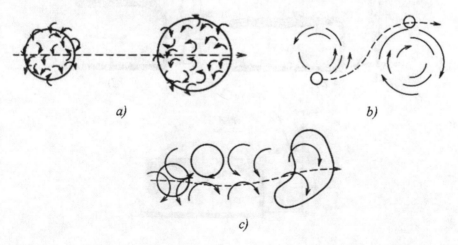

a)   b)

c)

Fig. 2 Air turbulence influence on impurities dispersion: **a**—a big cloud in homogenous field of small clubs; **b**—a small cloud in big field of clubs; **c**—a cloud in field of clubs with correlated sizes

Temperature of impurities released from pipes into the atmosphere can be higher than temperature environment. In such case it changes shape of the torch, curves its axis and increases the height reached by the emissions. Torch shape in the wind direction is largely dependent on turbulence structure. Atmosphere is dominated by large eddies caused by free convection with strong instability in the boundary layer. Size of the vortices is larger than the diameter of the torch. Their impact is mostly raise or lowering of the torch, giving it a wavy shape. These perturbations move along with the airflow and increase in amplitude. Torch with impurities contained in it reaches surface near the source where high concentration of contamination immediately occurs with such unstable transfer (Fig. 3a). Smoothed distribution of impurity concentration on underlying surface becomes like a fan after prolonged action of source due to the constant change of wind direction.

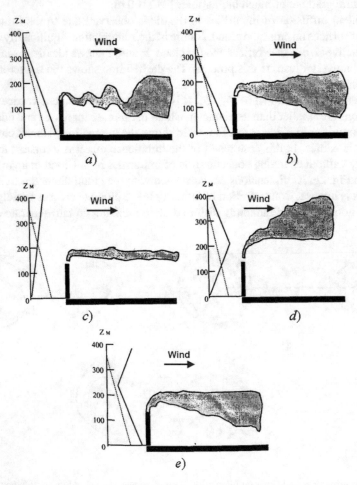

Fig. 3 Typical forms of the torch at different atmosphere stratification. Gradients of temperature: $\cdots$ dry adiabatic; —actual

Turbulent exchange is carried out mainly by small eddies of forced convection generated by the shift of wind speed in windy weather in the case of thermal stratification close to neutral. Therefore extension of torch from raised source occurs approximately with the same intensity both vertically and horizontally. So it takes form of symmetrical to the central axis cone (Fig. 3b). Diameter of the cone increases to downstream stream only due to turbulent diffusion. Vertical transfer is weakened by neutral thermal stratification. Torch reaches underlying surface at great distances from the source.

Turbulence is very weak and any rising ascending air flows are actively suppressed in the case of persistent atmospheric stratification. The torch remains thin due to the underdevelopment of vertical diffusion. It happened due to instability of the wind direction and it takes V-shape (it looks like a fan when viewed from above). The torch is in the form of a ribbon and can be kept so without significant changes up to distances of about 100 km from the source (Fig. 3c) in the case of a steady wind direction. The concentration of impurities on the underlying surface at this distance is almost zero in the absence of significant changes in the relief on the leeward side. It happened because there is no developed vertical transfer.

The most favorable condition for the dispersion of impurities is the location of the source above the inversion layer from the environmental point of view. Stably stratified air impedes the spread of impurities into the layers below. Moderately unstable stratification of the upper layers promotes diffusion of upward pollution (Fig. 3d). The opposite effect occurs if the inversion layer is located above the source and prevents the impurity from spreading to the upper layers. Air is stratified shakily from below. So, convective mixing promotes the impurity spreading to the underlying surface and increases concentration of pollution along the line corresponding to the torch axis (Fig. 3c) [19].

Also dispersion of impurities in the atmosphere and the level of surface concentration are affected by fogs, precipitation and solar radiation.

Air pollution increases under the influence of fog. Fog drops absorb harmful substances both near the surface and from the above polluted air layers. So, the concentration of impurities greatly increases in the layer of fog and decreases above it.

Precipitation plays an important role in the self-purification process of the atmosphere. Raindrops or snowflakes capture dust particles and carry them to the ground. Heavy precipitations make atmosphere cleaner. However, precipitation becomes a source of contamination of soil and surface water by harmful substances.

Solar radiation causes photochemical reactions in the atmosphere with formation of various secondary products. They often have more toxic properties than substances emitted by pollution sources.

Also, dispertion can change physical properties of impurities may due to coagulation.

In a real atmosphere industrial emissions are exposed to entire set of meteorological factors. It determines one or another level of pollution. Combination of meteorological conditions that contribute to impurities accumulation in the atmosphere is commonly referred to meteorological potential of atmospheric pollution. Combination of dispersion factors is atmospheric dispersion ability. Impact of different components of potential depends on the sources location, the emission parameters, and the frequency of its components. More frequent repeat of adverse conditions leads to increased accumulation of impurities and higher average pollution level.

Repeat of conditions favorable to impurities dispersion varies significantly from year to year. Variability of the impurity concentration due to changes in meteorological conditions can be very significant depending on the type of sources and the nature of their location throughout the city. Role of meteorological conditions in formation of average level of pollution can sometimes exceed role of quantity and composition of emissions [20].

Fast emission. Practically important tasks include study of pollutants propagation in the atmosphere during their fast emission. Fast emissions can be observed both in normal operation mode of technogenic object and during emergencies—fires and explosions. Distinctive feature of pollutants emissions is their decentralization and non-stationary propagation process when large quantities of harmful substances are emitted into the atmosphere over a short period of time. These emissions are physically close to instant sources.

Cloud of the dust and gas mixture rises to a certain height due to the action of an instant source. Air movement causes both dispersion and orderly transfer. Atmospheric vortices disperse the impurity if they are smaller in size than cloud, if larger—tolerate. Cloud is blurred (deformed) both in the direction of movement, and in the transverse and vertical directions due to turbulent exchange under the influence of wind. Wind direction determines trajectory. Wind speed is rate of movement of the impurity cloud from the emission site. Vertical temperature gradient at the cloud level is significant because unstable stratification promotes cloud elevation and delays it.

Impurity in cloud can be washed away by precipitation. Also it can be changed physically or chemically due to contact with moisture or other impurities in the atmosphere.

Generally accidental discharge can be chaotic mixture of all kinds of solids, liquids and gases. Solids and liquids size in diameter can be less than 0.1 microns and can reach several cm. Since particles with diameters greater than 10 microns are too large they fall off quickly after being released into the atmosphere. So, they are near the accident site. Particles with a diameter less than 1 micron remain in a suspended state for longer. Their propagation is more dependent on weather conditions. They can remain in the air for several days and be transported by wind over considerable distances [19].

5 Conclusions

According to the research results there are following the most determining factors influencing concentration distribution of pollutants in the air due to emissions from potentially dangerous technogenic objects are: type and conditions of emission, type of source (point, linear, site), wind direction and velocity, state of the atmosphere, chemical interaction with other substances in the air, gravitational deposition, gravitational sedimentation, absorption by underlying surface, surface roughness, terrain.

It is established that level of surface concentration of pollutants increases in case of increase of emission source power, density and particle size of the impurity, air temperature. It decreases with increasing of height and radius of emission source orifice, discharge temperature, speed of impurity emission from the source. Dependence of pollutant concentration on the wind speed is nonlinear.

The obtained results are the first stage of development of information and technical methods to prevent emergencies related to chemical contamination air in the territories of PDOs. It will be used in future for implementation of the second stage—development of mathematical models for pollutants propagation in the air from PDOs emissions during the corresponding emergency.

References

1. Wei, G., Sheng, Z.: Image quality assessment for intelligent emergency application based on deep neural network. J. Vis. Commun. Image Represent. **63**, 102581 (2019). https://doi.org/10.1016/j.jvcir.2019.102581
2. Wang, L., Wang, Y.-M., Martínez, L.: A group decision method based on prospect theory for emergency situations. Inform. Sci. **418–419**, 119–135 (2017). https://doi.org/10.1016/j.ins.2017.07.037
3. Islam, M.S., Ahmed, M.M., Islam, S.: A conceptual system architecture for countering the civilian unmanned aerial vehicles threat to nuclear facilities. Int. J. Crit. Infrastruct. Prot. **23**, 139–149 (2018). https://doi.org/10.1016/j.ijcip.2018.10.003
4. Ma, G., Wu, Z. BIM-based building fire emergency management: Combining building users' behavior decisions. *Automat. Constr.*, 2020, vol. 109. https://doi.org/10.1016/j.autcon.2019.102975
5. Fathi, R., Thom, D., Koch, S., Ertl, T., Fiedrich, F.: VOST: a case study in voluntary digital participation for collaborative emergency management. Inf. Process. Manage. 102174 (2019). https://doi.org/10.1016/j.ipm.2019.102174
6. Vallejo, D., Castro-Schez, J.J., Glez-Morcillo, C., Albusac, J.: Multi-agent architecture for information retrieval and intelligent monitoring by UAVs in known environments affected by catastrophes. Eng. Appl. Artif. Intell. **87**, 103243 (2020). https://doi.org/10.1016/j.engappai.2019.103243
7. Gong, P., Tang, X.B., Huang, X., Wang, P., Wen, L.S., Zhu, X.X., Zhou, C.: Locating lost radioactive sources using a UAV radiation monitoring system. Appl. Radiat. Isot. **150**, 1–13 (2019). https://doi.org/10.1016/j.apradiso.2019.04.037
8. Zabulonov, Y.L., Burtnyak, V.M., Zolkin, I.O.: Airborne gamma spectrometric survey in the Chernobyl exclusion zone based on oktokopter UAV type. Prob. Atomic Sci. Technol. **5**, 163–167 (2015)
9. Pereira, M.N.A., Schirru, R., Gomes, K.J., Cunha, J.L.: Development of a mobile dose prediction system based on artificial neural networks for NPP emergencies with radioactive material releases. Ann. Nucl. Energy **105**, 219–225 (2017). https://doi.org/10.1016/j.anucene.2017.03.017
10. Popov, O., Iatsyshyn, A., Kovach, V., Artemchuk, V., Taraduda, D., Sobyna, V., Sokolov, D., Dement, M., Yatsyshyn, T., Matvieieva, I.: Analysis of possible causes of NPP emergencies to

minimize risk of their occurrence. Nucl. Radiat. Saf. **81**(1), 75–80 (2019). https://doi.org/10.32918/nrs.2019.1(81).13

11. Holla, K., Moricova, V.: Specifics of monitoring and analysing emergencies in information systems. Transp. Res. Proc. **40**, 1343–1348 (2019). https://doi.org/10.1016/j.trpro.2019.07.186

12. Ropero, F., Vaquerizo-Hdez, D., Muñoz, P., Barrero, D.F., R-Moreno, M.D.: LARES An AI-based teleassistance system for emergency home monitoring. Cogn. Syst. Res. **56**, 213–222 (2019). https://doi.org/10.1016/j.cogsys.2019.03.019

13. Raja Shekhar, S.S., Venkata Srinivas, C., Rakesh, P.T., Deepu, R., Prasada Rao, P.V.V., Baskaran, R., Venkatraman, B.: Online Nuclear Emergency Response System (ONERS) for consequence assessment and decision support in the early phase of nuclear accidents—Simulations for postulated events and methodology validation. Prog. Nucl. Energy **119** (2020). https://doi.org/10.1016/j.pnucene.2019.103177

14. Tsiouri, V., Kovalets, I., Kakosimos, K.E., Andronopoulos, S., Bartzis, J.G.: Evaluation of advanced emergency response methodologies to estimate the unknown source characteristics of the hazardous material within urban environments. In: HARMO 2014—16th International Conference on Harmonisation within Atmospheric Dispersion Modelling for Regulatory Purposes, Proceedings, pp. 561–565 (2014)

15. Kovalets, I.V., Robertson, L., Persson, C., Didkivska, S.N., Ievdin, I.A., Trybushnyi, D.: Calculation of the far range atmospheric transport of radionuclides after the Fukushima accident with the atmospheric dispersion model MATCH of the JRODOS system. Int. J. Environ. Pollut. **54**(2–4), 101–109 (2014). https://doi.org/10.1504/IJEP.2014.065110

16. Cui, J., Lang, J., Chen, T., Cheng, S., Li, Y.: Emergency monitoring layout method for sudden air pollution accidents based on a dispersion model, fuzzy evaluation, and post-optimality analysis. Atmos. Environ. **222** (2020). https://doi.org/10.1016/j.atmosenv.2019.117124

17. Singh, R.K., Rao, A.R.: Steam leak detection in advance reactors via acoustics method. Nucl. Eng. Des. **241**(7), 2448–2454 (2011). https://doi.org/10.1016/j.nucengdes.2011.04.028

18. Babak, V.P., Babak, S.V., Myslovych, M.V., Zaporozhets, A.O., Zvaritch, V.M.: Principles of Construction of Systems for Diagnosing the Energy Equipment. In: Diagnostic Systems For Energy Equipments. Studies in Systems, Decision and Control, vol. 281, pp. 1–22. Springer, Cham (2020). https://doi.org/10.1007/978-3-030-44443-3_1

19. Popov, O., Yatsyshyn, A.: Mathematical tools to assess soil contamination by deposition of technogenic emissions. In: Dent, D., Dmytruk, Y. (eds.) Soil Science Working for a Living: Applications of soil science to present-day problems, pp. 127–137. Springer, Cham (2017). https://doi.org/10.1007/978-3-319-45417-7_11

20. Shkitsa, L.E., Yatsyshyn, T.M., Popov, A.A., Artemchuk, V.A.: The development of mathematical tools for ecological safe of atmosfere on the drilling well area. Neftyanoe khozyaystvo—Oil Industry, vol. 11, pp. 136–140 (2013)

21. Popov, O., Iatsyshyn, A., Kovach, V., Artemchuk, V., Taraduda, D., Sobyna, V., Sokolov, D., Dement, M., Yatsyshyn, T.: Conceptual approaches for development of informational and analytical expert system for assessing the NPP impact on the environment. Nucl. Radiat. Saf. **79**(3), 56–65 (2018). https://doi.org/10.32918/nrs.2018.3(79).09

22. Zaporozhets, A.O., Redko, O.O., Babak, V.P., Eremenko, V.S., Mokiychuk, V.M.: Method of indirect measurement of oxygen concentration in the air. Naukovyi Visnyk Natsionalnoho Hirnychoho Universytetu **5**, 105–114 (2018). https://doi.org/10.29202/nvngu/2018-5/14

23. Kovach, V., Lysychenko, G.: Toxic Soil Contamination and Its Mitigation in Ukraine. In: Dent, D., Dmytruk, Y. (eds.) Soil Science Working for a Living. Springer, Cham (2017). https://doi.org/10.1007/978-3-319-45417-7_18

24. Mergner, R., Janssen, R., Kovach, V., et al.: Fostering sustainable feedstock production for advanced biofuels on underutilised land in Europe. In European Biomass Conference and Exhibition Proceedings, pp. 125–130 (2017)

25. Yatsyshyn, T., Mykhailiuk, Y., Liakh, M., Mykhailiuk, I., Savyk, V., Dobrovolsky, I.: EStablishing the dependence of pollutant concentration on operational conditions at facilities of an oilandgas complex. East. Eur. J. Ent. Technol. **2**(10–92), 56–63 (2018). https://doi.org/10.15587/1729-4061.2018.126624

26. Shkitsa, L., Yatsyshyn, T., Lyakh, M., Sydorenko, O.: Means of atmospheric air pollution reduction during drilling wells. In: IOP Conference Series: Materials Science and Engineering, vol. 144 (2016). https://doi.org/10.1088/1757-899x/144/1/012009
27. Popov, O.O., Iatsyshyn, A.V., Kovach, V.O., Artemchuk, V.O., Kameneva, I.P., Taraduda, D.V., Sobyna, V.O., Sokolov, D.L., Dement, M.O., Yatsyshyn, T.M.: Risk assessment for the population of Kyiv, Ukraine as a result of atmospheric air pollution. J. Health Poll. **10**(25), 200303 (2020). https://doi.org/10.5696/2156-9614-10.25.200303
28. Popov, O., Iatsyshyn, A., Kovach, V., Artemchuk, V., Taraduda, D., Sobyna, V., Sokolov, D., Dement, M., Hurkovskyi, V., Nikolaiev, K., Yatsyshyn, T., Dimitriieva, D.: Physical features of pollutants spread in the air during the emergency at NPPs. Nucl. Radiat. Saf. **84**(4), 88–98 (2019). https://doi.org/10.32918/nrs.2019.4(84).11
29. Berland, M.: Modern Problems of Atmospheric Diffusion and Air Pollution. Gidrometeoizdat, Russia, Leningrad (1975)
30. Atmosphere: Handbook, Leningrad: Gidrometeoizdat, Russia (1991)
31. Fedotov, A.V. Analysis of methods for assessing and monitoring the environmental and economic consequences of emergency situations. Gornyy informatsionno-analiticheskiy byulleten **5**, 194–198 (2008)
32. Monin, A.S., Yaglom, A.M.: Statistical hydromechanics. Theory of Turbulence. St. Petersburg: Gidrometeoizdat, Russia (1992)

Investigation of Biotechnogenic System Formed by Long-Term Impact of Oil Extraction Objects

Teodoziia Yatsyshyn [ID], Nataliia Glibovytska [ID], Lesya Skitsa [ID], Mykhailo Liakh [ID], and Sofiia Kachala [ID]

Abstract The actual problem of the pollutants spread on the territory of oil and gas wells' influence and their effect on biocenoses is considered. The analysis of research aimed to study the role of the environmental oil pollution is conducted. The approach to conduct the research characterized by multifactoriality is carried out. The sources and causes of oil fluxes at different stages of the wells' life cycle are identified. The interrelations of pollutants migration in the biotechnological system "oil and gas well—environment" are established. Three oil wells of the Ivano-Frankivsk National Technical University of Oil and Gas (Ukraine) characterized by different external conditions of operation are identified as research objects. The conditions of long-term pollution formation on the oil wells territory are determined and the equipment condition that causes the influx of pollutants into the environment is analyzed. The sources of oil emissions from the equipment elements are identified. The species composition and life status of the biocenoses located on the territory of the wells are investigated. The leaves lesions of two species—Salix caprea L., Urtica dioica L. caused by pests are found in conditions of low oil pollution. It is found that low concentrations of petroleum in the environment stimulate the development of plants pests, while the high content of pollutants in the environment leads to their disappearance. The dominance of black ant Lacius niger L. in the area of high levels oil pollution and the displacement of other insect populations by this species is established.

Keywords Ash-slag dump · Ash-slag waste · Migration model · Environmental pollution · Energy facilities · Thermal power plant · TPP

T. Yatsyshyn (✉) · N. Glibovytska · L. Skitsa · M. Liakh · S. Kachala
Ivano-Frankivsk National Technical University of Oil and Gas, Ivano-Frankivsk, Ukraine
e-mail: yatsyshyn.t@gmail.com

V. Babak et al. (eds.), *Systems, Decision and Control in Energy I*, Studies in Systems, Decision and Control 298, https://doi.org/10.1007/978-3-030-48583-2_11

1 Introduction

Oil and gas production facilities pose a potential environmental hazard, both due to the presence of aggressive chemicals throughout the life cycle and because of the large amount of these facilities in all regions of the globe. Increasing volumes of hydrocarbon production and geological exploration of new fields remain a priority area of socio-economic development in Ukraine, despite the development of alternative energy sources. In western Ukraine oil and gas facilities are located near natural recreational zones of national importance, which creates a high risk of technogenic disturbance of valuable territories. As the main negative impacts are located on the surface of the atmosphere and surface groundwater, the areas of pollutants distribution can become transboundary. The pollution dynamics spread necessitates to study the potential and long-term effects of fluid pollution on the biota. The results of the research provide an opportunity to develop a system for of these territories reconstraction and recommendations for improving the environmental safety of technological processes in the life cycle of oil and gas wells life cycle.

2 Literature Analysis and Problem Statement

Petroleum environmental pollution is at the forefront of global ecological issues of today. According to the 2008/50/EC Directive of the European Parliament and of the Council it is necessary to reduce pollution to the levels of minimizing the hazardous impact on human health and the environment in general [1]. Environmental safety of oil and gas complex objects is a safe environment for any object functioning on the one hand, and no environmental impact on the object on the other [2]. However, such processes occurring at the wells create the conditions for the permanent entry of harmful substances into the environment.

In particular, studies by J. Min [3], H.-J. Kim [3], A. Oudin [4], P. Tuccella [5], J.L. Thomas [5], and others [6–8] indicate a significant environmental risk posed by atmospheric air pollution. The accumulation of soil contamination was studied in [9–14]. In particular, C.O. Adenipekun and I.O. Fasidi the *Lentinus subnudus* ability to mineralize soil contaminated with various concentrations of crude oil has been tested [12]. As a result of the experiment, the following indicators were obtained (Table 1).

The pernicious effects of the oil components on living organisms are known, which are manifested in disease occurrence, growth retardation, development and even death [15–17]. Polycyclic aromatic hydrocarbons having carcinogenic and mutagenic properties are particularly dangerous [18, 19]. At the same time, there are species of flora and fauna that are resistant to oil. In particular, some herbs and woody plants are capable of detoxifying toxic aromatic and aliphatic hydrocarbons to safe metabolism products. Thus, *Medicago sativa* L., *Betula pendula Roth.*, *Juglans regia* L., *Helianthus tuberosus* L., *Elodea canadensis* L. belong to persistent environmental

Table 1 TPH of crude-oil contaminated soil with *Lentinus subnudus* [12]

TPH (Mg/Kg)			
Treatment (concentration of crude oil) (%)	0 Months	3 Months	6 Months
1	9423 ± 0.05	6310 ± 0.03	3712 ± 0.01
2.5	34664 ± 0.10	15019 ± 0.20	7538 ± 0.10
5	62241 ± 0.20	52996 ± 0.18	8937 ± 0.30
10	140305 ± 0.15	104849 ± 0.25	14677 ± 0.20
20	256980 ± 0.25	198980 ± 0.25	12535 ± 0.13
40	336930 ± 0.40	285532 ± 0.30	16447 ± 0.25

remedies [20–25]. It is also known that low concentrations of oil in the environment stimulate the growth and development of individual organisms [26, 27].

Phytoremediation in anthropogenically-transformed ecosystems remediation shere is one of the most promising and environmentally sound methods in modern practical ecology [19, 28, 29]. The issue of plants adaptation to the oil components action and the species survival at different levels of ecosystem pollution is urgent. Investigations into the species composition of phytocenoses that have existed in oil-contaminated areas for a long time will help address the issue of combating environmental contamination.

3 Purpose and Objectives of the Study

The aim of the research is to study the "biocenosis—oil and gas well" system under the conditions of different levels of oil pollution exposure and the choice of approaches for their balanced interaction, which would provide optimal conditions of interaction in the area of the production facility influence, while preserving the natural resource and ecological potential of natural systems and their ability to self-regulate and recover.

4 Research Methods

Theoretical studies are performed by the method of synthesis and analysis of oil and gas complex objects life cycle impact on the biocenoses. Practical study of the territory under conditions of the oil and gas complex objects long-term impact is carried out by visual assessment and comparative analysis. The study is conducted under the conditions of three wells of Ivano-Frankivsk National Technical University of Oil and Gas (Ukraine), localized in the territory of the Dolyna region, which differed at the degree of oil pollution. Varieties of plants and insects distributed in the territory of oil production have been established.

5 Research Results

5.1 Choosing an Approach to Study the State of the Environment Within the Long-Term Exposure of Oil and Gas Objects

Oil and gas production facilities have a multifaceted environmental impact over the course of their lifecycle. Each stage of the cycle is characterized by scale, nature, duration, intensity of exposure etc. In order to identify the most appropriate approaches to prevent the negative effects of the oil and gas production process, it is important to consider all major factors that pose the environmental hazards. In this case, it is advisable to use a systematic methodology, which makes it possible to explore the object as a whole set of elements in the set of relations and relationships between them.

The state of the environment within the location of oil and gas production facilities is subject to intense impacts on different environments (Fig. 1). Impacts occur under a variety of factors, causing varying degrees of environmental risk. Therefore, the "oil and gas production facility (oil and gas well)—the environment" is a complex multifactorial system with diverse internal and external links, the study and association of which gives the opportunity to show a single picture and identifies the necessary ways to influence the system for increasing the environmental safety level.

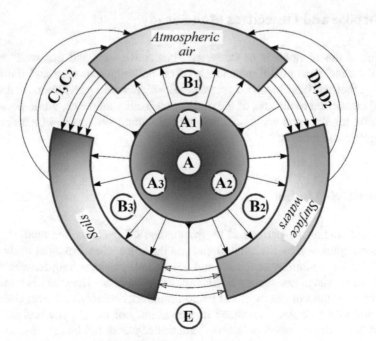

Fig. 1 Scheme of oil and gas facilities interaction dynamics with the environment

A—sources of negative impact (SNI); A1—SNI on the atmosphere; A2—SNI in water areas; A3—SNI on the soil cover; B1, B2, B3 are the ways of migration of pollutants from the SNI, respectively, into the atmospheric air, surface water and soils; C1, C2—secondary migration of pollutants from the soil surface into the atmosphere, by evaporation or combustion; D1, D2—secondary migration of pollutants from the water surface into the atmosphere, by evaporation or combustion; E is the secondary migration of pollutants in the soil-surface water system.

In order to study such a system, it is necessary to determine a sufficient number of informative indicators for adequate description at each level and stage of the life cycle. An important criterion for the selection and limitation of the study area is the determination of the size of the area where potential environmental risk is present (for instance in emergencies) and the allocation of ecosystems that are constantly influenced by regulated work to determine conceptual relationships.

5.2 Conditions of Oil Well Contamination Formation of the Territories

Oil and gas production processes are characterized by a multicomponent composition of working media (drilling fluids, extracted fluids and other process fluids) that can contain toxic components and can be flammable [30]. These factors, when summed up, create the risk of surrounding areas being affected.

The Fig. 2 shows the approximate duration of the oil and gas well life cycle stages, which traces the longest stage of exploitation.

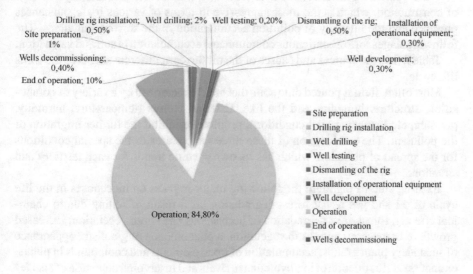

Fig. 2 Estimated duration of oil and gas well lifecycle stages

Table 2 Sources and causes of petroleum fluxes at different stages of the wells life cycle

No	Stage of the well lifecycle		Sources of oil fluxes	Causes of oil fluxes
1	Boring		Drilling mud, sewage	Emergencies and technological processes (lifting operations, borehole washing)
2	Preparation for operation	Testing	Oil reservoir, highly mineralized reservoir water, drilling fluids, drilling waste water	Technological processes (release of fluids into barns)
		Familiarization	Reservoir oil, reservoir of highly mineralized water	Technological processes
3	Operation		Formed highly mineralized water	Emergency situations, repair of equipment, technological operations
4	Operation end		Reservoir oil, reservoir of highly mineralized water	Emergency (old equipment)
5	Wells decommissioning		Reservoir oil, reservoir of highly mineralized water	Emergency situations (destruction of cement bridges, tectonic activity)

When evaluating the entire life cycle of a well, it is necessary to identify the stage of construction, which is the most aggressive in terms of various toxic substances presence and the intensity of pollution accumulation in the surrounding areas. The following stages form a continuous contaminants accumulation in soil and vegetation.

Table 2 lists the sources and causes of the oil fluxes at different stages of the wells life cycle.

Most often, fluid is poured onto soils that are characterized by a variety of composition, structure, humidity, and the like. External factors (temperature, humidity, pressure, etc.) create special conditions, promote or inhibit the further migration of the pollutant. The combination of these three factors creates the special conditions for the spread of pollution, which has its own specific trends for each territory and situation.

Existing studies highlight the following major impacts on biocenoses in the life cycle of oil and gas wells: forest degradation (as a result of felling due to chemical effects); forest fires; degradation of herbaceous and shrub vegetation; increased growth of herbaceous and shrub vegetation, a phenomenon of giantism; appearance of secondary plant groups; accumulation of toxic elements and compounds in plants; extinction and extinction of ichthyofauna in rivers and reservoirs; depletion of species composition and reduction of birds and mammals.

a   b

Fig. 3 Layout of "64-Valley" well site: **a**—playground arrangement; **b**—installation of deep-water rod pump with hydraulic drive

5.3 Practical Studies of Oil Well Impact Areas

At Ivano-Frankivsk National Technical University of Oil and Gas (IFNTUOG), which is the leader in providing high-level technical and non-technical training services to Ukraine's energy sector—a major driver of the Ukrainian economy is a research pilot site where 5 wells are located.

Ivano-Frankivsk National Technical University of Oil and Gas (IFNTUOG) scientists conducted a study of the biotechnogenic system formed on the territory of long-term impact of oilfield objects located at the landfill.

The landfill consists of five oil wells that were drilled in the period 1970-1991. (age of the wells: 64-Valley—51 years; 81-Vygoda-Vitvitsa—30 years; 152-North Valley—46 years; 61-North Valley—48 years; 1-Smolyan—28 years) and are located at different fields, which allows to carry out ecological scientific and educational researches in different conditions, taking into account the features of the topography, the coefficient of ecological stability, vegetation, microclimate and so on. The studies were conducted on three wells, which are currently at the final stages of operation.

Within the range of 64-Valley the installation of a deep-driven hydraulic, which is currently the only one in Ukraine, allows to study its technical advantages from the point of view of environmental safety in comparison with the used rocking machines and to study it optimal operating modes (Fig. 3) [31]. The characteristic odor of aromatic hydrocarbons in the landfill site indicates the presence of gaseous toxic compounds in the air.

"152-North Valley" well is characterized by the highest level of environmental pollution, which is explained by the strong leakage of oil to the soil surface of the territory due to the failure of the protective equipment of the well (Fig. 4). Due to the long leakage of oil into the adjacent area near the well, a lake about 10 meters long was formed.

Fig. 4 "152-North Valley" well

The "81-Vygoda-Vitvitsa" well is characterized by low oil flow into the environment and is marked by the presence of oil spots up to 1.5 m (Fig. 5).

The species composition of phytocenoses on the territory of the three wells is almost the same. Herbaceous plants dominate—*Carex hirta* L., *Matricaria recutita* L., *Urtica dioica* L., *Cirsium arvense* L., *Ranunculus acris* L., *Equisetum arvense* L.,

Fig. 5 «81-Vygoda-Vitvitsa» well

Fig. 6 Urtica dioica L. pests existing in the environment of oil pollution

Rosa multiflora L., *Medicago sativa* L., *Taraxacum officinale* L., *Trifolium pretense* L., *Plantago major* L. Species diversity of plant communities in different conditions of oil pollution is almost the same. Plants well adapted to the effects of oil grow comfortably with varying degrees of contamination. However, among the studied plants, nettle and willow drooping are characterized by a sensitivity to the effects of petroleum products, which is manifested in the appearance of plant pests. In this case, leaf pests strike explored plants in the zone of light pollution ("81-Vygoda-Vitvitsa" and 1-Smolyan) (Figs. 6 and 7).

This is due to the fact that low concentrations of oil in the environment stimulate the growth and development of some insects, including plant pests. At the same time, persistent insect species displace other insect species that cannot adapt to pollution.

Also interesting is the fact that in the conditions of severe pollution—"152- North Valley"—there is a prosperity of the population of black ants—*Lacius niger* L. of the genus *Formica*, which displaces other species of ants from their territory. The black ant population is significantly decreasing in the area of low level of oil pollution ("81-Vygoda-Vitvitsa" and "64-Valley"), which indicates the ecological confinement of the species to extreme conditions of existence.

6   Discussion of the Results

The conducted researches allow to estimate the biotechnogenic system created by long-term accumulation of oil pollution, both in soils and through migration through air. The condition of the well equipment that is on the surface can be assessed

Fig. 7 Salix caprea L. pests existing in the environment of oil pollution

visually, however, well equipment that is in an aggressive environment needs separate monitoring. As the accumulation of pollutants is also possible through flows in the pressurized wellbore.

A stable ecosystem has been formed on the territory of the research experimental site, which includes oil-resistant plant remedies that are known in the literature for their ability to detoxify petroleum components. The uniqueness of the landfill is precisely the ability to study in the environment the totality of life processes of plants associated with the decomposition of petroleum components in their organisms. The study at monitoring points of the effect of different concentrations of petroleum products on remedial plants allows more effective use of phyto-objects in landscaping for the recovery of oil-contaminated environment. In addition, further studies on the species composition of insects in the landfill allows to analyze the possible adaptive responses of class members, including plant pests, to the effects of oil and value in the trophic network.

7 Conclusions

The conditions of pollution formation on the territory of oil and gas wells are analyzed. The dynamics of the interaction of oil and gas production facilities with the environment is determined and the duration of the stages of the life cycle of the wells is analyzed. In situ studies of oil wells are conducted on the territory of the research experimental landfill, which allows to evaluate the condition of the equipment on the well floor, to determine the sources of pollutants and the reaction of biocenosis of the area.

Stable phytocenoses are formed under conditions of different levels of oil pollution of the environment, resistant to adverse conditions of existence. Weak levels of oil pollution lead to the appearance of pests of individual plants—Urtica dioica L. and Salix caprea L., which are vulnerable to pests that thrive under low concentrations of petroleum products. The high level of environmental oil pollution determines the dominance of black ants in biocenosis and is indicated by the absence of plant pests that exist in these conditions.

References

1. Directive 2008/50/EC of the European Parliament and of the Council of 21 May 2008 on ambient air quality and cleaner air for Europe (OJ L 152, 11.6.2008, pp. 1–44)
2. Yatsyshyn, T., Mykhailiuk, Y., Liakh, M., Mykhailiuk, I., Savyk, V., Dobrovolskyi, I.: Establishing the dependence of pollutant concentration on operational conditions at facilities of an oil-and-gas complex. East. Eur. J. Ent. Technol. **2**(10–92), 56–63 (2018). https://doi.org/10.15587/1729-4061.2018.126624
3. Min, J., Kim, H.-J., Min, K.: Long-term exposure to air pollution and the risk of suicide death: a population-based cohort study. Sci. Total Environ. **628–629**, 573–579 (2018). https://doi.org/10.1016/j.scitotenv.2018.02.011
4. Oudin, A.: Short review: Air pollution, noise and lack of greenness as risk factors for Alzheimer's disease- epidemiologic and experimental evidence. Neurochem. Int. (2020). https://doi.org/10.1016/j.neuint.2019.104646
5. Tuccella, P., Thomas, J.L., Law, K.S., Raut, J.-C., Marelle, L., Roiger, A., Weinzierl, B., Denier van der Gon, H.A.C., Schlager, H. and Onishi, T.: Air pollution impacts due to petroleum extraction in the Norwegian Sea during the ACCESS aircraft campaign. Elem. Sci. Anth. **5**, 25 (2017). https://doi.org/10.1525/elementa.124
6. Hendryx, M., Luo, J., Chojenta, C., Byles, J.E.: Air pollution exposures from multiple point sources and risk of incident chronic obstructive pulmonary disease (COPD) and asthma. Environ. Res. **179**, 108783 (2019). https://doi.org/10.1016/j.envres.2019.108783
7. Popov, O.O., Iatsyshyn, A.V., Kovach, V.O., Artemchuk, V.O., Kameneva, I.P., Taraduda, D.V., Sobyna, V.O., Sokolov, D.L., Dement, M.O., Yatsyshyn, T.M.: Risk assessment for the population of Kyiv, Ukraine as a result of atmospheric air pollution. J. Health Poll. **10**(25), 200303 (2020). https://doi.org/10.5696/2156-9614-10.25.200303
8. Shkitsa, L.E., Yatsyshyn, T.M., Popov, A.A., Artemchuk, V.A.: The development of mathematical tools for ecological safe of atmosfere on the drilling well area. Neftyanoe khozyaystvo—Oil Industry, **11**, 136–140 (2013)
9. Khamehchiyana, M., Charkhabib, A.H., Tajika, M.: Effects of crude oil contamination on geotechnical properties of clayey and sandy soils. Eng. Geol. **89**(3–4), 220–296 (2007). https://doi.org/10.1016/j.enggeo.2006.10.009
10. Kovach, V., Lysychenko, G.: Toxic soil contamination and its mitigation in Ukraine. In: Dent, D., Dmytruk, Y. (eds.) Soil Science Working for a Living. Springer, Cham, (2017). https://doi.org/10.1007/978-3-319-45417-7_18
11. Mazurek, R., Kowalska, J.B., Gąsiorek, M., Zadrożny, P., Wieczorek, J.: Pollution indices as comprehensive tools for evaluation of the accumulation and provenance of potentially toxic elements in soils in Ojców National Park. J. Geochem. Explor. **201**, 13–30 (2019). https://doi.org/10.1016/j.gexplo.2019.03.001
12. Adenipekun, C.O., Fasidi, I.O.: Bioremediation of oil-polluted soil by Lentinus subnudus, a Nigerian white-rot fungus. Afr. J. Biotechnol. **4**(8), 796–798 (2005)

13. Popov, O., Yatsyshyn, A.: Mathematical tools to assess soil contamination by deposition of technogenic emissions. In: Dent, D., Dmytruk, Y. (eds.) Soil Science Working for a Living: Applications of Soil Science to Present-Day Problems, pp. 127–137. Springer, Cham (2017). https://doi.org/10.1007/978-3-319-45417-7_11

14. Pedroso, A., Bussotti, F., Papini, A., Tani, C., Domingos, M.: Pollution emissions from a petrochemical complex and other environmental stressors induce structural and ultrastructural damage in leaves of a biosensor tree species from the Atlantic Rain Forest. Ecol. Indic. **67**, 215–226 (2016). https://doi.org/10.1016/j.ecolind.2016.02.054

15. Kaur, N., Erickson, T., Ball, A., Ryan, M.: A review of germination and early growth as a proxy for plant fitness under petrogenic contamination—knowledge gaps and recommendations. Sci. Total Environ. **603–604**, 728–744 (2017). https://doi.org/10.1016/j.scitotenv.2017.02.179

16. Glibovytska, N.I., Karavanovych, K.B.: Morphological and physiological parameters of woody plants under conditions of environmental oil pollution. Ukrainian J. Ecol. **8**(3), 322–327 (2018)

17. Tran, T., Mayzlish, E., Eshel, A., Winters, G.: Germination, physiological and biochemical responses of acacia seedlings (Acacia raddiana and Acacia tortilis) to petroleum contaminated soils. Env. Poll. **234**, 642–655 (2018). https://doi.org/10.1016/j.envpol.2017.11.067

18. Gouda, A., El-Gendy, A., Abd El-Razek, T., El-Kassa, H.: Evaluation of Phytoremediation and Bioremediation for Sandy Soil Contaminated with Petroleum Hydrocarbons. IJESD **7**(7), 490–493 (2016). https://doi.org/10.18178/ijesd.2016.7.7.826

19. Shevchyk, L.Z., Romanyuk, O.I.: Analysis of biological methods of recovery of oil-contaminated soils. Sci. J. Sci. Biol. Sci. **1**(4), 31–39 (2017)

20. Lewis, J., Qvarfort, U., Sjöström, J.: Betula pendula: a promising candidate for phytoremediation of TCE in northern climates. Int. J. Phytoremed. **17**(1), 9–15 (2015). https://doi.org/10.1080/15226514.2013.828012

21. Minoui, S., Shahriari, M., Minai, D.: Phytoremediation of crude oil-contaminated soil by Medicago sativa (Alfalfa) and the effect of oil on its growth. Phytoremed. Green Energy, 123–129. (2015) https://doi.org/10.1007/978-94-007-7887-0_8

22. Ikeura, H., Kawasaki, Yu., Kaimi, E., Nishiwaki, J., Noborio, K., Tamaki, M.: Screening of plants for phytoremediation of oil-contaminated soil. Int. J. Phytoremed. **18**, 460–466 (2016). https://doi.org/10.1080/15226514.2015.1115957

23. Marchand, C., Hogland, W., Kaczala, F., Jani, Y., Marchand, L., Augustsson, A., Hijri, M.: Effect of Medicago sativa L. and compost on organic and inorganic pollutant removal from a mixed contaminated soil and risk assessment using ecotoxicological tests. Int. J. Phytoremed. **18**(11), 1136–1147 (2016). https://doi.org/10.1080/15226514.2016.1186594

24. Jahantab, E., Jafari, M., Motasharezadeh, B., Ali, T.: Remediation of petroleum-contaminated soils using Stipagrostis plumosa, Calotropis procera L., and Medicago sativa under different organic amendment treatments. ECOPERSIA **6**(2), 101–109 (2018)

25. Glibovytska, N., Karavanovych, K., Kachala, T.: Prospects of phytoremediation and phytoindication of oil-contaminated soils with the help of energy plants. J. Ecol. Eng. **20**(7), 147–154 (2019). https://doi.org/10.12911/22998993/109875

26. Courchesne, F., Turmel, M., Cloutier-Hurteau, B., Constantineau, S., Munro, L., Labrecque, M.: Phytoextraction of soil trace elements by willow during a phytoremediation trial in Southern Québec. Canada. Int. J. Phytoremediation **19**(6), 545–554 (2017). https://doi.org/10.1080/15226514.2016.1267700

27. Yergeau, E., Tremblay, J., Joly, S., Labrecque, M., Maynard, C., Pitre, F., St-Arnaud, M., Greer, C.: Soil contamination alters the willow root and rhizosphere metatranscriptome and the root–rhizosphere interactome. The ISME Journal **12**, 869–884 (2018). https://doi.org/10.1038/s41396-017-0018-4

28. Lim, M.W., Lau, E.V., Poh, P.E.: A comprehensive guide of remediation technologies for oil contaminated soil—present works and future directions. Mar. Pollut. Bull. **109**(1), 619–620 (2016). https://doi.org/10.1016/j.marpolbul.2016.04.023

29. Panchenko, L., Muratova, A., Turkovskaya, O.: Comparison of the phytoremediation potentials of Medicago falcata L. and Medicago sativa L. in aged oil-sludge-contaminated soil. Environ. Sci. Pollut. Res. **24**(3), 3117–3130 (2017). https://doi.org/10.1007/s11356-016-8025-y

30. Shkitsa, L., Yatsyshyn, T., Lyakh, M., Sydorenko, O.: Means of atmospheric air pollution reduction during drilling wells. In: IOP Conference Series: Materials Science and Engineering, vol. 144, (2016). https://doi.org/10.1088/1757-899x/144/1/012009
31. Velychkovych, A.S., Panevnyk, D.O.: Study of the stress state of the downhole jet pump housing. Naukovyi Visnyk Natsionalnoho Hirnychoho Universytetu **5**, 50–55 (2017)

Biological Risk of Aviation Fuel Supply

Iryna Shkilniuk⬤ and Sergii Boichenko⬤

Abstract The identification of hazards is a prerequisite for managing risk factors for flight safety. Fuel is the blood of an aircraft. The mass of fueled jet fuel is up to 70% of the maximum take-off weight of modern aircraft. The main link in all civil aviation activities is flight safety. Many documents ICAO, IATA and Joint Inspection Group focuses on pollution fuels. Danger is an inherent attribute of aviation activity. But their manifestations and possible consequences can be prevented by applying different strategies of compensatory measures, designed to contain the potential of hazards that could lead to the creation of unsafe ones operating conditions. There are many risks in the production of fuels, including the risk of microbiological contamination. The biological risk factor can be defined as biological matter capable of self-replication and which can have a destruction effect on the fuel. Today it is known 200 species of microorganisms, including 30 families that can use hydrocarbons as sole source of carbon and energy. These include bacteria, yeast and fungi. Microorganisms have the selective ability related to various hydrocarbons, and this ability is determined not only by the difference in the structure of substance, and even the number of carbon atoms that are the part of their structure. The places of microbiological colonies development of on the fuel life cycle are established during the analysis of biological risk of aviation fuel supply. The risk of uncontrolled microbial contamination is generally greatest in tropical regions. The results of research have shown that fuel for jet engines with microbiological contamination has increased corrosion properties. The corrosion is one of the most dangerous types of metal destruction of aircraft structures. Over 50% of all corrosion processes due to the influence of microorganisms. The conclusion that biocorrosion of the fuel system and aircraft structures is part of the problem fuel with microbiological contamination are made and justified in these materials. A host of problems will likely surface when uncontrolled microbial growth is allowed to develop. Microbial activity has been shown to cause degradation of fuel hydrocarbons. Flight safety also will likely be compromised, as well as increased maintenance and cost. Not all microorganisms, however, cause the same problems.

I. Shkilniuk (✉) · S. Boichenko
National Aviation University, Kyiv, Ukraine
e-mail: i_shkilniuk@ukr.net

© The Editor(s) (if applicable) and The Author(s), under exclusive license   179
to Springer Nature Switzerland AG 2020
V. Babak et al. (eds.), *Systems, Decision and Control in Energy I*, Studies in Systems,
Decision and Control 298, https://doi.org/10.1007/978-3-030-48583-2_12

180 I. Shkilniuk and S. Boichenko

Keywords Aviation fuel · Biological risk · Microbiological contamination · Fuel system · Biocorrossion · Biofilm · Acidity · Biocomponent · Hydrocarbon · Biocide

1 Introduction

Aviation fuel supply exists as much as aviation is more than a hundred years. Fuel is the blood of an aircraft. The mass of fueled jet fuel is up to 70% of the maximum take-off weight of modern aircraft. The main link in all civil aviation activities is flight safety. The most important condition for ensuring the safety of flights is the use of aircraft in the range of expected operating conditions, taking into account operational limitations established in the norms of airworthiness.

Now ICAO is disturbed formed by the world tendency of entering of contaminated aviation fuel in airport. Many documents ICAO, IATA and Joint Inspection Group focuses on pollution fuels. ICAO issued a directive DOC 9977 "Guide to the supply of aviation fuel in civil aviation" and IATA issued EI/JIG STANDARD 1530 "Quality assurance requirements for the manufacture, storage and distribution of aviation fuels to airports". The essence of these documents is that all parties involved are jointly responsible for ensuring the quality, purity and possibility of quality control at every stage of production, supply and operation of aviation fuel.

Fuel and air machinery loss during operation are major aircraft losses for modern aircraft. The main factors and parameters that determine the indicated losses in operation are flight path, speed and altitude; equipment reliability; fuel conditioning.

The study of theoretical and practical aspects of risk, its analysis and assessment is becoming increasingly relevant, because the risk in today's economic environment has a significant impact on the results of enterprises.

2 Literature Data Analysis and Problem Statement

The active study on microbial growth in the composition of petroleum fuels began in the USA during the creation of jet aircraft.

During the creation of jet aviation in the USA, began active study of questions connected with microorganisms' development in oil fuels. The work on this question in our country mainly was to determine fuels biostability in laboratory conditions. Purposeful researches of fuels biostability in operating conditions were not conducted practically [1].

In 1956, the United States Air Force recognized that its widely-used JP-4 fuels were microbial contaminated when Air Force B-47 and KC-97 flight operations were affected. Two years later, a B-52 crash was directly attributed to clogging of fuel system screens and filters by some form of fuel contamination. In that same year, the Wright Air Development Center determined that sludge accumulation in tanks used to store kerosene-type fuels was a common occurrence [3].

More instances of contamination and corrosion surfaced in the late 1950's and early 1960's and reached near epidemic proportions in storage tanks and aircraft fuel cells at various locations. At the beginning of 1962, approximately 52 governmental and non-governmental agencies were involved in various phases of research on microbiological contamination of fuels.

The modern world legislation raises the level of requirements for quality aviation fuels. In 2012, the International Civil Aviation Organization has developed directive 9977/AN 489 'Guidelines for the supply of aviation fuel for civil aviation', which focused on the clean air fuels, including microbiological contamination.

3   Main Material

Danger is an inherent attribute of aviation activity. But their manifestations and possible consequences can be prevented by applying different strategies of compensatory measures, designed to contain the potential of hazards that could lead to the creation of unsafe ones operating conditions

Air transport is a major consumer of high-quality fuels and lubricants. For large scale high oil consumption issue efficiency of aviation technology, economy and management of aviation fuel has an important public and economic value. The efficiency and reliability of the fuel system software greatly depends on the quality of aviation fuel. The largest number of failures and malfunctions elements of the fuel system, engine and aircraft related to fuel quality and purity.

The introduction of ICAO and IATA standards, satisfaction of safety and operational safety requirements, economic indicators, financial profitability require a risk-based approach in aviation fuel supply [2, 3].

The risk indicator is introduced for the quantitative characteristics of the safety of objects. Risk is a measure of danger [4]. The analysis of the threats to sustainable aviation fuel supply is to identify all sources of threat and assess their impact on flight safety. A flight safety risk factor is the predicted probability and severity of an impact or outcome caused by an existing hazard or situation.

The process of performing risk analysis has consisted of the following consecutive procedures:

– to determine the risk factor for flight safety;
– the probability of flight safety risk factors;
– the severity of flight safety risk factors;
– airworthiness risk tolerance;
– management of flight safety risk factors.

The key stages in the risk analysis process are the identification of risks and their classification. At the risk identification stage, the risks that exist at different stages of the jet fuel cycle life were analyzed [5].

Stages of the fuel life cycle:

1. oil recovery stage;
2. oil refining stage;
3. stage of production of commodity jet fuel;
4. stage of jet fuel transportation;
5. jet fuel storage stage;
6. fuel use stage.

Specific aspects can be distinguished from the point of view of the chemmotological reliability. One of them is the purity of fuel, the presence or absence of mechanical impurities, water, microorganisms and other contaminants that should not be present in the fuel when shipped from production sites, but which can accumulate during transportation, storage, pumping and other operations.

There are many risks in the production of fuels, including the risk of microbiological contamination. The hydrocarbon component is the most dangerous raw material from this point of view. On the one hand, this is the basis of fuels, on the other is the source of potential infection of microbiological oil destructors.

Destruction of materials usually occurs under the influence of not a single group of microorganisms and an entire complex including bacteria and fungi. One group of microorganisms of its own activity prepares a substrate for another. The process is very complex and is due to a large number of factor. The substrate is of paramount importance, it forms the formation of such substrates new, functionally interconnected units, as a microbial association or biocenosis.

Microbes may be introduced into fuels as products cool in refinery tanks. Bacteria and fungi are carried along with dust particles and water droplets through tank vents. In seawater ballasted tanks, microbes are transported with the ballast. Vessel compartments ballasted with fresh, brackish, or seawater, all of which may contain substantial numbers of microbes, may easily become contaminated with the microbes transported with the ballast water [6].

Today it is known 200 species of microorganisms, including 30 families that can use hydrocarbons as sole source of carbon and energy (Fig. 1). These include bacteria, yeast and fungi (Table 1).

Active development of the fuel and the fuel systems of microscopic fungi (*Hormoconis resinae, Penicillium, Aspergillus fumigatus, Paecilomyces variotii,* etc.) recognized the most dangerous. Fungi form a dense mycelium, the accumulation of which not only clog pipelines and fuel filters, but also create numerous localized areas of corrosion on the surfaces of fuel systems. A most active destructor of aviation fuel until recently was recognized *Cladosporium resinae* (modern name *Hormoconis resinae)* or *Amorphoteca resinae.* This so-called "kerosene" fungi [7, 9]. Today this group is extended with fungi *Monascus floridanus,* which is inherent in the ability to develop rapidly in the aviation fuel [4]. Fungi have some morphological, physiological and genetic features, good with which they occupy the dominant position among organisms causing biological damage.

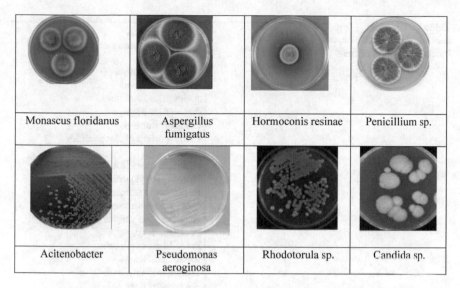

| Monascus floridanus | Aspergillus fumigatus | Hormoconis resinae | Penicillium sp. |
| Acitenobacter | Pseudomonas aeroginosa | Rhodotorula sp. | Candida sp. |

Fig. 1 The main strains of microorganisms-petroleum product destroyers

The first stable products of n-alkanes oxidation are the primary alcohols. The next is usual biological conversion of alcohols to aldehydes and aldehyde to acid [8]. The general scheme of reactions:

$$R\text{-}CH_2\text{-}CH_3 + [O] \rightarrow R\text{-}CH_2\text{-}CH_2OH\text{-}2H \rightarrow R\text{-}CH_2\text{-}CHO\text{-}2H + HOH \rightarrow R\text{-}CH_2\text{-}COOH.$$

Microorganisms have the selective ability related to various hydrocarbons, and this ability is determined not only by the difference in the structure of substance, and even the number of carbon atoms that are the part of their structure.

It is proved that microbial contamination of fuel is connected to microbiological enzymatic oxidation of hydrocarbons with formation of organic acids that have surface active properties. The speed and depth of the microbial oxidation of aviation fuel depend on their carbohydrate composition. Hydrocarbons with a linear structure of the molecules are destroyed faster than their branched isomers. Aliphatic hydrocarbons (paraffin's) are less biostable than aromatic. Therefore, fuels that contain mostly paraffin hydrocarbons can be destroyed by microorganisms faster than those containing more aromatic compounds. Cycloalkanes are more difficult to microbiologically destruction than alkanes, due to the presence of a cyclic structure that is heavier than oxidation. Strains that are capable of biodegradation of cyclic alkanes include bacteria of the genera *Cordonia, Xanthobacter*, and others. Seams that are capable of biodegradation of cycloalkanes have specific enzyme systems that are different from the enzyme systems used by microorganisms to oxidize non-cyclic alkanes.

The research of activity of growth of active and potential destructors spent in fuel by the value of the accumulation of biomass after a month of cultivation (Fig. 2).

Table 1 The main microorganisms, that cause biocontamination of fuels	Fungi	Acremonium sp.
		Altenaria altenarata
		Aspergillus sp.
		Aspergillus clavatus
		Aspergillus flavus
		Aspergillus fumigatus
		Aspergillus niger
		Cladosporium sp.
		Cladosporium cladosporoides
		Fusarium sp.
		Fusarium moniliforme
		Fusarium oxysporum
		Hormoconis resinae
		Monascus floridanus
		Paecilomyces variotii
		Penicillium sp.
		Penicillium cyclioium
		Rhinocladiella sp.
		Trihoderma viride
		Trichosporon sp.
	Bacteria	Acitenobacter
		Alcaligehes
		Bacillus sp.
		Clostridium Sporogenes
		Flavobacterium difissum
		Micrococcus sp.
		Pseudomonas sp.
		Pseudomonas aeroginosa
		Serratia marcescens
	Yeasts	Candida sp
		Candida famata
		Candida guilliermondii
		Candida lipolytica
		Rhodotorula sp.

The greatest value of biomass is defined for the *Hormoconis resinae* isolated from the tank of the aircraft [10].

The ability to grow potential destructors for various hydrocarbons was studied. Micromycetes grow on all tested liquid hydrocarbons, except for hexane [9]. The tendency to increase in hydrocarbons with more long carbon chains (C_{10}-C_{17}) was observed in *Hormoconis resinae* and *Monascus floridanus*. The greatest importance of biomass increase in *Hormoconis resinae* was observed on heptadecane ($C_{17}H_{36}$), *Monascus floridanus* on hexadecan ($C_{16}H_{34}$).

After appearance in fuel tanks, microorganisms may either stick to overhead surfaces or settle through the product. Some microbes will adhere to tank walls, whereas others will settle to the fuel/water interface (Fig. 3). Most growth and activity takes place where fuel and water meet. The tank bottom fuel/water interface is the most obvious fuel/water boundary. However, there is also a considerable area of

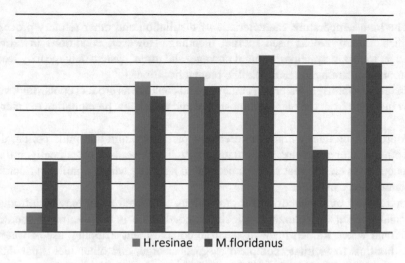

Fig. 2 Biomass of active destructors after a month growth in hydrocarbons, g/100ml

Fig. 3 Schematic of Fuel Tank Bottom Sample with Significant Microbial Contamination and Biodeterioration

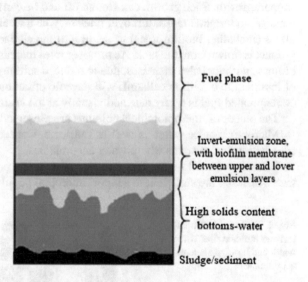

fuel/water interface on the interior surface of tank-shells. Microorganisms require water for growth. Although bacteria and fungi can be present in the fuel phase, their growth and activity is restricted to the water phase of fuel systems. The water phase includes volumes ranging from trace (several μL) to bulk (>1 m^3) accumulations and water entrained within deposits that accumulate on system surfaces. Typically, fuel and system deterioration is caused by the net activity of complex microbial communities living within slimy layers called biofilms. Biofilms may be found on tank roofs, shells, at the fuel/water interface, and within bottom sludge/sediment [6].

The high temperature characteristic of distillation and other refinery processes sterilize refinery stocks used in fuel blending. However, conditions in refinery tankage, transport systems, terminal tankage, and users' system tankage may lead to microbial contamination and possible biodeterioration.

In refinery tankage, water can condense and coalesce as product cools. Tank vents draw moisture from the outside atmosphere and may allow precipitation to enter the tank.

Moreover, product withdrawal creates a partial vacuum that pulls pollen, dust, and other microbe-carrying particulates through tank vents. Consequently, refinery products tanks are the first stage of petroleum handling where significant microbial contamination can occur.

In transport by means of tanker or pipeline, additional water may be introduced by condensation. In contrast to pipelines, condensate is not the major source of additional water. Rather, inadequate cargo compartment stripping, use of water as false bottoms to facilitate complete cargo discharge, and other incidental, intentional water use provide substantial water to fuel tanks. Biofilms can form on tanker or pipeline surfaces where they entrain water, inorganic particles, and nutrients to support growth. Such growth can slough off and be carried to terminal and end user tankage. In terminal tanks, turnover rates may be a week or longer, allowing particulates (including biofilm flocs) to settle into the sludge and sediment zone before product is drawn from the tank. As turnover rates increase, the likelihood of drawing biomass with fuel also increases, due to reduced settling times. Population densities of less than two million cells/mL will have no effect on fuel clarity. Consequently, contaminated fuel is rarely detected visually at the terminal rack.

The places of microbiological colonies are presented in Fig. 4 [6].

Microbes require water as well as nutrients. Consequently, they concentrate at sites within fuel systems where water accumulates.

Water is essential for microorganisms' growth and proliferation. Even negligible traces of water are sufficient to support microbial populations.

Fig. 4 Fuel supply scheme (arrows indicate sites where water and biologicals tend to accumulate)

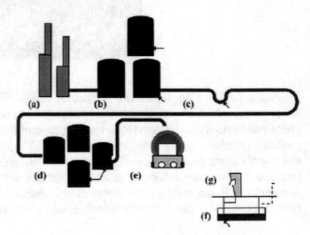

Nutrients are divided into macronutrients and micronutrients. Carbon, hydrogen, oxygen, nitrogen, sulfur, and phosphorus (CHONSP) comprise the macronutrients, and most of these are readily available in fuels. Only phosphorous is likely to be growth limiting in most fuel systems. A variety of elements, including calcium, sodium, potassium, iron, magnesium, manganese, copper, cobalt, nickel, and other metals, are required in trace quantities. None of these elements is limiting in fuel systems. Fuel systems that provide both the requisite water and nutrients will support microbial growth and proliferation.

The rate of microbial growth increases with increasing temperature within the physiological range (temperature range within which growth occurs) of a given microorganism. Microbes are generally classified into three groups, based on their temperature preferences/requirements. Some microbes require low temperatures (<20 °C). Others thrive in superheated environments (>100 °C). However, the physiological range of the microbes most commonly recovered from fuel tanks is 0 to 35 °C, with growth optimal between 25 and 35 °C.

(a) refinery distillation towers
(b) refinery product tanks
(c) fuel transportation pipeline (low points in pipeline trap water)
(d) distribution terminal tanks
(e) commercial dispensing rack and tank truck
(f) retail/fleet underground storage tank
(g) retail/fleet dispensing system.

The some strains of *Hormoconis resinae* are capable of developing in a fuel at a temperature of 50 °C and the strains of *Aspergillus fumigatus* survive in aviation kerosene up to 80 °C. The growth of *Hormoconis resinae* in aviation fuel is fixed at a temperature of 28 °C [10]. The activity of fungi decreases with increasing or decreasing temperature (Table 2).

Table 2 The growth of fungi in fuel at positive temperatures (in points)

Fungi	The time the manifestation of signs of growth, days	The temperature, °C			
		9	18	28	36
Hormoconis resinae	7	0	0	0	0
	14	0	1	2	0
	21	1	2	3	1
Phialofora sp	7	0	0	2	0
	14	1	2	4	2
	21	2	2	4	3

0—no signs of growth, 1—turbidity of the water layer, the formation of precipitation, 2—the appearance of large flakes in the water layer, 4—the formation of small clots, 5—the formation of large clots

The risk of uncontrolled microbial contamination is generally greatest in tropical regions. However, in the absence of adequate housekeeping practices, microbial contamination problems can also occur in fuel systems located in cold climates.

Water pH is generally not a controlling factor in fuel systems. Most contaminant microbes can tolerate pH's ranging from 5.5 to 8.0. As with temperature, there are microbes that prefer acidic environments (some grow in the equivalent of 2 N sulfuric acid) and others that grow in alkaline systems with pH > 11. Fuel tank bottom-water pH is usually between 6 and 9 [6, 8].

As water activity tends to be greatest at interface zones, this is where microbes are most likely to establish communities, or biofilms. Numbers of microbes within biofilms are typically orders or magnitude greater than elsewhere in fuel systems. Biofilms can form on tank overheads, at the bulk-fuel, bottom-water interface, and on all system surfaces.

Using fuel hydrocarbon vapors as their carbon source, microorganisms can colonize tank overheads, where condensation provides the necessary water activity. Biofilms on overheads generally look like slimy stalactites.

Whereas a 1-mm thick biofilm on a tank wall may seem negligible, it is 100 times the thickness of most fungi, and 500 to 1000 times the longest dimension of most bacteria. This seemingly thin film provides a large reservoir for microbial activity. Within the biofilm micro-environment, conditions can be dramatically different from those in the bulk product.

Microorganisms consortia (communities) give the biofilm community characteristics that cannot be predicted from analysis of its individual members.

Microorganisms are able to consume hydrocarbons directly excrete waste products that other consortium members use as food. The net effect is a change in pH, oxidation-reduction (or redox) potential, water activity, and nutrient composition that has little resemblance to the environment outside the biofilm [6].

Microbes growing anaerobically produce low molecular weight organic acids (formate, acetate, lactate, pyruvate, and others). These acids accelerate the corrosion process by chemically etching the metal surface. There are data demonstrating that biofilm communities can deplasticize the polymers used in fiberglass synthesis. Such activity can result in catastrophic tank failure and is most likely to occur along the longitudinal centerline (the same place of the greatest frequency of MIC pinholes).

Biosurfactants facilitate water transport into the fuel phase and some fuel additive partitioning into the water phase. Other metabolites may accelerate fuel polymerization. Produced at concentrations that are difficult to detect against the complex chemistry of fuel components, these metabolites can have a significant deleterious effect on fuel stability. Although most of the change occurs within a few centimeters of the biofilm-fuel interface, product mixing can distribute metabolites throughout the fuel system.

More particularly after microbiological contamination of aviation fuels the following effects are observed in the presence of the above-mentioned favorable conditions [9]:

- *change in physical and chemical properties of fuels*, namely increasing of major physical-chemical parameters values as acidity, kinematic viscosity, refractive index, pH, content of actual resins and others. Also characteristic features are the formation of sediment, turbidity fuel and peculiar odor;
- *corrosion of storage tanks for aviation fuels*. Corrosion development of bottom part where accumulates water sludge, especially on verge of system distribution "fuel-water", corrosive damage of aircraft tanks, corrosion of aircraft power constructions;
- *clogging and damage of fuel filters, pumps and fuel systems*. Sedimentation of mycelium and bacteria colonies at the inner walls of the fuel systems leads to clogging of pipelines, filters, pumps and fuel systems;
- *threat to the safety of aircrafts flights*. Changing the physical, chemical and exploitation properties of aviation fuels leads to early clogging of filters, pollution of regulating equipment, causing unstable operation of the fuel system, and therefore can cause failure of the engine, and even complete failure of the system, and as a consequences is appearance of accidents and emergency landings.

The changing the acidity of fuel is an important consequence of the proliferation of microorganisms (Fig. 5). Acidity of fuel due to the presence of organic or inorganic acids or their derivatives [11]. Therefore, the change in pH value of the fuel can indirectly determine the dynamics of microbial growth in fuel. The acidity of the fuel due to their content of organic acids and acidic compound. This quality index

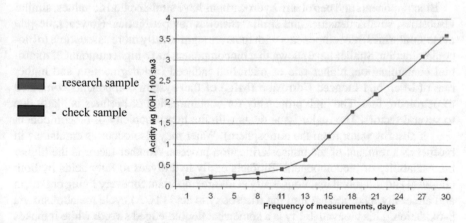

■ - research sample

■ - check sample

Fig. 5 Change jet fuel acidity under the influence of microbiological contamination

characterizes the presence of the fuel products that provoke an increase in the rate of wear and corrosion of engine friction pairs and the air supply system, as well as corrosion of tanks, pipes and fittings. The metabolic products of microorganisms destructors fuels contain organic acids and increase the acidity of fuels.

The aviation industry is experiencing increasing pressure from the public and environmentalists who say that the increase in traffic volume and number of aircraft operated by causing serious damage to the environment, besides aviation is one of the largest consumers of fuel and lubricants. Necessary to replace petroleum hydrocarbons alternative raw materials. Carbon biomass carbon cheaper oil. However, the conversion of the "cheap" carbon in the consumer goods while expensive. Before commissioning, the risk and the effect on reliability of equipment should be evaluated.

Biocomponents is a materials derived from the lipids of plants or animals, which can be used directly in existing combustion engines. In a base catalyzed process, triglycerides are broken down through the transesterification of the ester bond linking the glycerol backbone with the fatty acids. Through methanolysis, glycerol is substituted with a methyl group to produce single chain fatty acid methyl esters (FAME). FAME are structurally similar to petroleum alkanes and furthermore have suitable physical and chemical properties, which allows for use it in engines [13, 14].

There are problems with biocomponents which need to be addressed before biocomponents become a fully viable alternative to fossil fuels. The problem is the higher propensity of biocomponents towards microbial contamination compared with petroleum hydrocarbon.

Biocomponents and petroleum hydrocarbons have similar calorific values, similar viscosities, similar densities and similar material compatibilities. However, despite these similarities, biocomponents with fuels are significantly more susceptible to biocontamination. Studies have shown that biocomponent has a higher amount of microbial contamination, higher rate of microbial induced fuel degradation and higher rate of Microbial Induced Corrosion (MIC) of fuel system components compared to petroleum fuel. The high propensity for contamination of biofuels is likely due to several factors. The major issue deals with the hygroscopicity of biocomponent (i.e., it absorbs water from the atmosphere). Water may also occur as emulsions in biofuel as a remnant of the transesterification process. Another factor is the higher bioavailability of biocomponent. Biofuels easily hydrolyzes to fatty acids by both chemical and microbial reactions. Fatty acids are important for every living organism and are easily incorporated into the tricarboxylic acid (TCA) cycle metabolism via β-oxidation. This bioavailability is a somewhat double-edged sword: while it makes use of biocomponent more difficult on a daily basis, biocomponent degrades in soil and water environments in a few days, diminishing the environmental impact of fuel spills.

Studies of microbiological stability of aviation fuels have been carried out in the laboratory of the Ukrainian Scientific-Research Center of Chemmotology and certification of fuel and technical liquids. Samples of fuels were submitted jet fuel TC-1 and automobile gasoline A-95 (Table 3). The methyl ester of fatty acids of sunflower oil was selected as the bio-component jet fuel TC-1. Gasoline A-95 has

Table 3 List of samples

Number of samples	Samples
1	Jet fuel TC-1
2	Jet fuel TC-1 + 10% biocomponent
3	Jet fuel TC-1 + 20% biocomponent
4	Jet fuel TC-1 + 30% biocomponent
5	A-95
6	A-95 + 5% ethanol
7	A-95 + 10% ethanol
8	A-95 + 15% ethanol
9	Biocomponent

been selected as an alternative to leaded aviation gasoline. Ethanol was added in different concentrations to gasoline [9].

The test specimens were infected colonies microorganisms and control samples are placed in a lit environment with ambient temperature 20–25 °C. The studies were conducted 2 weeks. Control of changes were made every week on visual and chemical methods. The acidity of fuel was chosen as the chemical method of control microbiological growths. The choice of rate is due to the mechanism of microbial degradation and metabolism products biodestruktors. The change of the value of the indicator will be judged on the impact biocontamination quality fuels, fuels and microbiological stability of the rate of growth of microorganisms.

The essence of the method lies in the titration of the acidic compounds of the test product with an alcoholic solution of potassium hydroxide in the presence of a color indicator. Acid number is expressed in mg KOH/100 cm^3. This is standart method [11]. Determination of the acid number performed in the semi-automatic titrator 702 SM Titrino Mettler Toledo.

Test results are presented in Fig. 6. The values on the vertical axis of the diagram of Fig. 1 is the acidity, expressed in mg of KOH/100 cm^3 of fuel.

Biocomponents such as the methyl ester of fatty acids of sunflower oil is subject to the same biodegradation by microorganisms like jet fuel.

Fig. 6 The change of acidity of traditional and alternative fuels under the influence of microbiological contamination

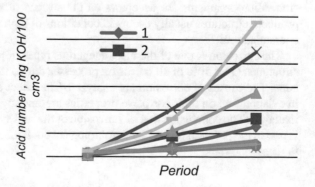

Table 4 List of samples for research and test results

Sample	The test result
Clean fuel for jet engines TC-1	
Clean fuel for jet engines Jet A-1	
Fuel for jet engines TC-1 with microbiological pollution	
Fuel for jet engines Jet A-1 with microbiological pollution	

Corrosive activity of aviation fuels is estimated by such indicators as acidity, tests on a copper plate, the content of water-soluble acids and alkalis, water content, sulfur content, etc. The test on a copper plate is a universal method for qualitative evaluation of corrosion activity of aviation motor fuels. Increased corrosion of fuels with microbiological contamination is confirmed by the results of experiments carried out in Test interactive laboratory "AviaTest" by the index of copper plate.

The gist of the method is as follows. The prepared copper plate is immersed in a certain amount of sample, heated and kept at a temperature of 100 °C. for 3 h. After this time, the plate is removed, washed and compared with standards of corrosion.

The list of samples and results of their tests is presented in Table 4.

The results of research have shown that fuel for jet engines with microbiological contamination has increased corrosion properties. The appearance of dark plaque and changes in color on the plates are very noticeable. The appearance of copper plates, which were kept in fuels for jet engines with signs of microbiological contamination, does not correspond to reference samples.

Corrosion occurs under the influence of the products of vital activity of microorganisms present in fuels. The fungi and many bacteria form ammonia, hydrogen sulfide, and various organic acids in the process of metabolism, most of which are characterized by high corrosive activity (Fig. 7, Table 5). In the course of its development, microorganisms destroy inhibitors that protect metal and stimulate its corrosion. Microorganisms are acceptors on the surface of metals. Corrosion of metal products, structures usually occurs in conditions of high humidity in the presence of contamination.

The corrosion is one of the most dangerous types of metal destruction of aircraft structures. Over 50% of all corrosion processes due to the influence of microorganisms. The causes of corrosion are analyzed. The effect of microbial contamination in aviation fuel on their corrosive properties are investigated and described in these materials. The conclusion that biocorrosion of the fuel system and aircraft structures is part of the problem fuel with microbiological contamination are made and justified in these materials.

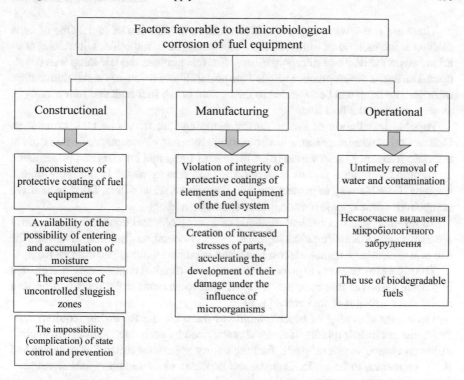

Fig. 7 The causes of microbiological corrosion

Risks	Principal Types of Microorganisms
Blockage of pipes, valves, filters	Fungi; polymer-producing bacteria
Increased water content	All
Sludge formation	All
Surfactant production	Fungi; and aerobic bacteria
Corrosion of storage tanks and lines	Fungi; and anaerobic bacteria
	All
Production of suspended solids in the fuel	Fungi; and aerobic bacteria
	All
Breakdown of hydrocarbons	Aerobic bacteria and fungi
Shortened filter life	Sulfur-reducing bacteria
Fouling injectors	(SRB)
Increased sulfur content of fuel	Undetermined
Shortened life of engine parts	Fungi
Penetration of protective tank	Endotoxing-producing
Health problem	bacteria, SRB

Table 5 Potential consequences caused by microorganisms in aviation fuel systems

There are many ways to prevent biological contamination of fuels. One of such method is the method of ultraviolet and electromagnetic radiation. Ultraviolet radiation causes the death of microorganisms. For this purpose the UV lamp was developed. During its development excluded the possibility of explosion and inflammation of fuels. The lamp can be mounted to the bottom of the fuel tank and move along it, as well as along the fuel line.

Possible installation of lamps during pumping fuel from one tank to another. Destruction of microorganisms is also possible by using electromagnetic radiation at a certain frequency radio waves [8]. Colonies of fungi and bacteria can be removed by filtration through a porous material, the pore size of which is not more than 2 microns. Possible way to protect the fuel through bacterial filters, filled with silver compounds (for example, cotton, glass, synthetic rubber).

To physical and mechanical methods of microbiological contamination control are also include centrifugation followed by agglomeration filtration, flotation, the use of ion-exchange resins, electro hydraulic deposition, ultrasonic control [6, 8].

The most effective way to protect the fuel from biological contamination at present is biocide additives that reduce activity of microorganisms in jet fuels and prevent biological corrosion of fuel tanks [1, 5].

During the choosing of biocide additives there are the following requirements: they must not impair quality of fuels, characterized by prolonged action, detrimental effect on engine structural parts, fuel regulatory apparatus, reliability of filters and filter separators, to be toxic. Combustion products of these substances should not cause adverse effects on the environment [4].

Biocide additives may be soluble in fuels, and water cushion and destroy microorganisms in both phases [8].

Many biocide products have been tested abroad that meet the above requirements, there are the following: ethyleneglycol monometyl ether and Biofora F [8].

Ethyleneglycol monometyl ether is anti water crystallization additive, with glycerol. However, it was found that glycerol actively contributes to the microorganisms, and without it ethyleneglycol monometyl ether reduces their growth. In addition to the fuel for air jet engines—0, 1–0, 15% by weight, substance concentrates in water up to 20%, which not only prevents the formation of ice crystals, but also reproduction of microorganisms.

Biofor F after the penetration to oil product is concentrated in the free water. The mechanism of this substance action is also based on increasing of osmotic pressure. The effectiveness of the substance is in its lower concentrations in the water. This additive has the following drawback: when added to jet fuel is deposited on the blades of aircraft turbines and can cause them to corrosion due to increased acidity of water.

Long-term monitoring of fuel tanks coated with furan resins showed that microorganisms in these tanks is reduced [8].

There is well-known antiwater crystallization liquid "I-M", which is a product of association ethyl cellosolve and methanol. Liquid "I-M" is designated for use as additives to the fuel for the air jet engines, refueled aircrafts of civil aviation to decrease the probability of icing aircrafts and helicopters filters at low temperatures.

We researched bactericidal properties of the additive that caused by containing of methanol [10].

There are used biocides that have the active components—cellosolve, compounds of nickel, copper and other metals, heterocyclic compounds in quantities 0,0001–0,005% [7].

Due to increasing the range of biocide additives, there were studied bactericidal activity of such compounds dimethyl-dialkil-ammonium chloride ($[R_2(CH_3)_2N]Cl$) and dimethyl-alkyl-benzyl-ammonium chloride ($[R(CH_3)_2NC_6H_5–CH_2]Cl$) for aviation fuels—gasoline and fuel TS-1 for air jet engines [8].

During the study of these compounds has been established [8] that the amount of 0, 05% or more above mentioned additives reduce the growth of all microorganisms in the aviation gasoline and fuel TS-1.

It was studied biocide activity of such compounds: zinc salts of synthetic fatty acids, mixed salts of zinc and mercury, acetic and oleic acids. With addition to jet fuel in concentrations of 0.05–0.1%, they found sufficient activity, reducing the number of microorganisms on 75–85%. The salts of higher carboxylic acids of chrome, copper and lead, and also naphthenate of iron, copper and chromium were low-toxic [8].

Taking into account problem actuality of protection from both fuels accumulation of static electricity, and from microbiological contamination, was obtained complex additive that has antibacterial and anti-static properties. Mixtures of bactericidal and anti-static additives of different composition were studied; both bactericidal components applicated dimethyl-dialkil-ammonium chloride [8]. Simultaneously, this additive is an effective anti-static additive in concentration of 0.003%, increases conductivity and reduces oil electrification during their motion [8].

It is set that the antiwater-crystallization additive PFA-55 MB has high bactericidal effect for jet engines. Addition to jet fuel in an amount of 0,05–0,15% of PFA-55 MB additive practically fully prevents development of microorganisms and corrosion of fuel tanks of jet engines. This additive is the most widespread abroad [9].

It was found that 8-hydroxyquinoline and disalicildenpropandiamin in addition to fuel for air jet engines brand TS-1 in concentration 0,2 and 0,1% diminished growth of microorganisms accordingly on 88 and 75%. Primary amines of $C_{12}–C_{15}$, which was added to the fuel in an amount of 1%, diminished growth of microorganisms on 95%.

Special experiments reflected that active biocide additives in the water-fuels systems there can be substances that do not dissolve in fuel, but soluble in water. Thus, the complete destruction of microorganisms in the environment in fuel TS-1 was observed when injected into the water phase one of the following substances: 0,04% 1,2-diaminopropana or hexamethyldiamin, 0,12% ethylendiamin, hydroxylamine of hydrochloric acid or methylamine tartrate, 0,16% trimethylamine or n-butylamine.

Growth of microorganisms reducing on 98% is observed when the content in the water phase 0,08% n-butylamine, etylendyamina, hydroxylamine hydrochloride or methylamine oxalic acid.

Inhibition of microorganisms increasing by 70, 75 and 90% was observed in environment of fuel TS-1 when in the water phase added respectively 0,24% chromium acetate, 0,16% chromium nitrate, 0,16% copper acetate [9, 10].

There is also known multifunctional additive IPOD (isopropyloktadetsylamin).

Bacteria fungicidity of additive on the base of gas condensates was studied. Unlike the other additives, it obtained from hydrocarbon fractions (145–280) °C of gas condensates. Adding of the additive in amount of 0,1% destroyed microorganisms within 10–15 days on 100%.

Synthesized additive has not only antibacterial, but also antioxidant and anti-corrosion properties. The additive addition to final concentration of 0,1% prevents sediments in fuel on 80% [5].

Katon FP 1.5 of the company ROHM AND HAAS (U.S.A.) is one of the highly effective biocides that used worldwide for various fuels. In the nomenclature of the International Union of Theoretical and Applied Chemistry, an active component of Katon FP 1.5 is defined as 5-chloro-2-methyl-4-isotyazolin-3-one.

Today many foreign companies producing biocide additives to petroleum prod-ucts, such as: «Bang and Bonsomer», «THOR», «ROHM AND HAAS» and others [8].

The authors conducted research on the efficiency of modern biocide additives (applications) of mentioned above foreign manufacturers (Table 6). The research was conducted by the method of diffusion zone, which is testing the microbiological stability of jet fuel protected by antimicrobial additives with different concentra-tions in the Petri dish on nutrient dry agar for cultivation of microorganisms. Zones diameter of growth absence characterized the degree of test fuel stability.

It was used a mixture of aerobic bacteria (*Pseudomonas, Bacterium, Mycobac-terium*) as a test cultures, allocated from the affected oil.

The research results of biological stability of aviation fuel RT, protected by biocide additives with the method of diffusion zone are shown in Fig. 8.

So, this diagram represents that the best antimicrobial properties has the following additives: GROTAN OX, AKTICIDE OX, AKTICIDE MV14.

A host of problems will likely surface when uncontrolled microbial growth is allowed to develop. Microbial activity has been shown to cause degradation of fuel hydrocarbons. Flight safety also will likely be compromised, as well as increased maintenance and cost. Not all microorganisms, however, cause the same problems.

Assessment or analysis of risk is a process for identifying hazards, assessing the probability of an event and its consequences. The ratio of risk objects and risky events makes it possible to determine the link between the biological risk in the field of the use of aviation fuel with technogenic and economic risks. Technogenic risk is a complex indicator of reliability of elements of technical means of operation. It expresses the probability of an accident or disaster during the operation of machines and mechanisms, in particular vehicles, and the implementation of technological processes. The source of Technogenic risk is the violation of the rules of operation of technical systems, the untimely conduct of preventive inspections. Economic risk is determined by the ratio of benefits and harm that society receives from a particular activity.

The authors of this work identified and systematized the consequences and the risks of microbiological contamination of aviation fuel (Fig. 9) [12–14].

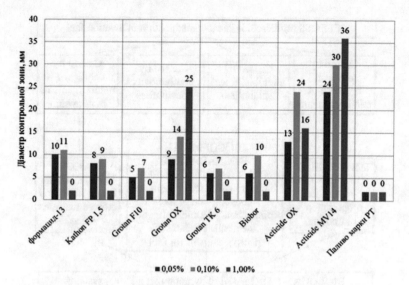

Fig. 8 Comparative distribution of research results of fuels biological stability, protected by biocide additives by diffusion zone method

Table 6 Results of the experiment by the method of zonal diffusion	Additive name	Zone diameter, mm		
		Additive concentration in fuel RT		
		1%	0,1%	0,05%
	Formacide	0	11	10
	KATHON	0	9	8
	Grotan F10	0	7	5
	Grotan OX	25	14	9
	Grotan TK 6	0	7	6
	ACTICIDE KL	0	10	6
	ACTICIDE OX	16	27	13
	ACTICIDE MV14	36	30	24
	Pure fuel RT(control)	0	0	0

Risk reduction is an action to reduce the likelihood of a negative event or mitigate the consequences of this event if it occurs.

A key factor is a multi-aspect approach to fuel hygiene to eliminate the inconveniences and costs associated with contamination of the fuel system. Each air operator should conduct his own risk assessment in order to determine the optimal regime.

An important component of this regime is the frequent checks of drainage systems, as well as regular testing and monitoring of microbiological contamination in the entire fuel system.

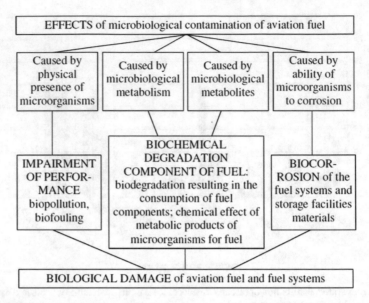

Fig. 9 The risks and consequences of microbial contamination of aviation fuels

References

1. Shkilniuk, I., Boychenko, S., Turchak, V.: The problems of biopollution with jet fuels and the way of achiving solution. Transport. S. Boychenko. **23**(3), C. 253–257 (2008)
2. DOC 9977. Guide to the supply of aviation fuel in civil aviation
3. Rauch, M.E., Graef, H.W., Rozenzhak, S.M., Jones, S.E., Bleckmann, C.A., Kruger, R.L., Naik, R.R., Stone, M.O.: Characterization of microbial contamination in United States Air Force aviation fuel tanks. J. Ind. Microbiol. Biotechnol. **33**, 29–36 (2006)
4. ISO 31000:2018 Risk management—Guidelines
5. Orel, C.M., Maliovaniy M.C.: Ризик. Basic concepts-Lviv: publishing house NU « Lvivska politechnika » , p. 88 (2008)
6. ASTM Standard D 6469 Guide for Microbial Contamination in Fuels and Fuel Systems. ASTM International
7. Aviation chemmotology: fuel for aviation engines. Theoretical and engineering bases of application: textbook. N.S. Кuliк, A. Ф. Aksionov, L. S. Yanovskiy, S. V. Boichenko—К.: НАУ, p. 560 (2015)
8. Shkilniuk, I., Boichenko, S.: Methodically organizational principles of biological stability providing of aviation fuel. Transactions of the Institute of Aviation of Warsaw **4** (237), 76–83 (2014)
9. Influence of microbiological contamination on acidity of traditional and alternative aviation fuels. Problems of chemmotology. Theory and practice of rational use of traditional and alternative fuel and lubricants: monograph. Sergii Boichenko, Kazimir Lejda, Vasiliy Matiychik, Petro Topilnitskiy.—К.: Center for Educational Literature, pp. 341–346 (2017)
10. Vasilieva, A.A., Chekunova, L.N., Poliakova, A.V.: Influence of temperature on growth and viability of Hormoconis resinae u Phialophora sp., Developing in aviation fuelsInfluence of temperature on growth and viability of Hormoconis resinae u Phialophora sp., developing in aviation fuels. Magazine Mycology and phytopathology, publishing house Izdatel'stva Nauka **43**(4), 312–316 (2017)
11. ASTM D3242.: Standard Test Method for Acidity in Aviation Turbine Fuel (2011)

12. Microbiological control in the system of civil aviation jet fuel supply. Iryna Shkilniuk, Sergii Boichenko, Kazimierz Lejda/ Proceedings of the 19th Conference for Junior Researchers 'Science—Future of Lithuania' TRANSPORT ENGINEERING AND MANAGEMENT, 6 May 2016, pp. 90–94 Vilnius, Lithuania
13. Iakovlieva, A.V., Boichenko, S.V., Vovk, O. O.: Overview of innovative technologies for aviation fuels production. Chem. Chem. Technol. **7**(3), 305–312 (2013)
14. Iakovlieva, A., Vovk, O., Boichenko, S., Lejda, K., Kuszewski, H.: Physical-chemical properties of jet fuels blends with components derived from rapeseed oil. Chem. Chem. Technol. **10**(4), 485–492 (2016)

An Improved Approach to Evaluation of the Efficiency of Energy Saving Measures Based on the Indicator of Products Total Energy Intensity

Olena Maliarenko⬤, Vitalii Horskyi⬤, Valentyna Stanytsina⬤, Olga Bogoslavska⬤, and Heorhii Kuts⬤

Abstract Existing methodological approaches to determining the total energy intensity of products and its components are presented. The methodological approach to determining the direct (technological) energy intensity of products on the example of combined production of energy at Combined Heat and Power (CHP) plants with the allocation of energy costs for certain stages of the technological process is improved: preparation of fuel, its submission to burners, burning in a boiler plant and the implementation of natural resources. This detailing of energy costs makes it possible to compare options for upgrading power equipment across all power plant workshops. Two types of CHP plant are considered: coal and gas, appropriate equipment structure, technologies that can be implemented to replace existing, less efficient ones. Three variants of possible modernization of the CHP plant on coal fuel and two—on natural gas are presented. The technological potential of energy saving in the case of implementation of the modern technologies for the production of energy at the coal CHP plant as more environmentally hazardous has been calculated, provided that the environmental requirements and the implementation of environmental measures are achieved. Effective technologies for reducing emissions of nitrogen oxides, sulfur oxides and particulates have been considered as environmental measures.

Keywords Energy · District heating · Energy supply · Energy efficiency · Direct energy intensity · Through technological energy intensity

1 Introduction

Combined generation of electricity and heat is a major trend in modern development of energy supply systems in the world. The share of electricity production at the CHP plants in Ukraine coincides with the share of district heating production in the G8 + 5 countries and is 11–19% [1]. In the EU countries, the heat utilization rate of the CHP plants reaches 75%, compared to 55% in Ukraine. The development of the latest

O. Maliarenko · V. Horskyi · V. Stanytsina (✉) · O. Bogoslavska · H. Kuts
Institute of General Energy of NAS of Ukraine, Kyiv, Ukraine
e-mail: st_v_v@hotmail.com

© The Editor(s) (if applicable) and The Author(s), under exclusive license to Springer Nature Switzerland AG 2020
V. Babak et al. (eds.), *Systems, Decision and Control in Energy I*, Studies in Systems, Decision and Control 298, https://doi.org/10.1007/978-3-030-48583-2_13

technologies indicates the widespread implementation of innovative solutions at the CHP plant with using renewable energy sources, including heat pumps, biomass, solid waste incineration, electric boilers that use excess electricity from wind farms and solar power station (SPS) with centralized heat storage.

The potential for the development of CHP plants in Ukraine is determined by the extensive infrastructure of the district heating system, powerful heating boilers, and backbone and distribution heat networks. Heating boilers with a thermal capacity exceeding 20 Gcal (23.26 MW) make up 2% of the total installed capacity of all heating boilers in Ukraine and produce about 60% of the total heat production. This is the potential for the introduction of combined electricity generation technology in cogeneration units.

In Ukraine, the combined production of heat and electricity at CHP plants, is carried out at 34 city utility power plants, with an installed power of 2856.23 MW and a heat capacity of 13466 Gcal/h, and at 87 thermal power plants of enterprises with an installed power of 2705 MW and a thermal capacity of 17897 Gcal/h [2], among which the largest installed capacity is the Kiev CHP-6 (750 MW), CHP-5 (700 MW) and Kharkiv CHP-5 (540 MW). In 2017, 6371 million kWh of electricity and 11571 thousand Gcal of thermal energy were produced at the city CHP plants. In 2017, the industrial CHP plants produced 4224 million kWh of electricity and 17522 thousand Gcal of thermal energy [2].

The structure of used fuel, according to the statistical report for 2017, is presented in Fig. 1 [3].

As can be seen from Fig. 1, natural gas (51.1%) occupies the largest share in the CHP plants, while coal (39%) occupy the second place.

Accordingly, CHPs that burn natural gas are large (500–2550 tonnes steam per hour), medium (200–420 tonnes steam per hour) and low capacity (less than 200

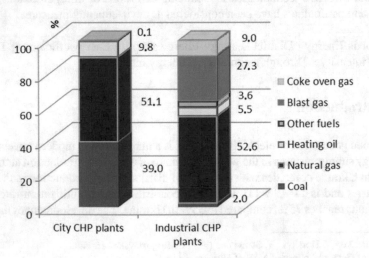

Fig. 1 Structure of fuel use for conversion to thermal and electric energy at different CHP plants in 2017 [3]

tonnes steam per hour). Coal CHP plants—medium and low capacity. The last are few: Sumy CHP plant (40 MW, 350 Gcal/h), Darnytsia CHP plant (180 MW, 1080 Gcal/h), Chernihiv CHP plant (210 MW, 500 Gcal/h), Kalush CHP plant (210 MW, 500 Gcal/h). All CHP plants have steam turbine technologies. Types of boiler installations are different in design (with U-, T-shaped and tower layout and combustion chamber), different types of burner devices are used [4, 5].

The share of natural gas in the structure of fuel use at industrial CHP plants (Fig. 1) is almost the same as for city CHPs (52.6%), but blast furnace gas (27.3%) occupies second place. Due to the greater variety of fuel used, boilers of industrial CHP plants differ in their burners and accessories.

Increase the energy efficiency of the main and auxiliary equipment of a CHP plant is an important task, since the National Emission Reduction Plan for large combustion plants includes the following CHP plants: Darnytsa, Dneprodzerzhinsk, Kramatorsk, Kremenchuk, Lviv, Mykolaiv, Odesa, Ohtyrka, Severodonctsk, Sumy, Kharkiv, Kherson, Cherkasy, Chernihiv, Shostka, Mariupol (2 CHP plants), Simferopol, Saky, Kamysh-Burun and industrial CPHs: Lisichans oil refinery, Alchevsk metallurgical plant (MP), Makiivka MP, Avdiivka coke and chemicals plant, Ilyich Iron and Steel Works of Mariupol, Sumy Machine-Building Science-and-Production Association, Pervomaysk "Energohimproekt", three CHP plants Kryvorizhstal, CHP plant Zaporizhstal, CHP plant DniproAzot, CHP plant PivdenMashbudzavod, CHP plant Azovstal. Reducing fuel use is the main direction for reducing greenhouse gas emissions [6].

The most commonly used energy efficiency indicators that can be used to evaluate the operation of a CHP plant are the utilization heat fuel factor, the energy conversion efficiency of the equipment by type of energy, unit energy recourses consumption by type, energy intensity of products (heat and electricity) and energy loss (fuel and energy carrier) [7]. Efficiency is usually defined for separate equipment (steam generator, turbine) or power plant as a whole as a result of multiplying of the efficiency of power installation along the technological system [8]. In determining most of these indicators, except utilization heat fuel factor, it is necessary to divide the total energy of the fuel to two energy products with different quality.

An important indicator of energy efficiency, that characterizes the complete technological cycle of production, is the total energy intensity of products [9], which allows to calculate the energy efficiency of replacement, modernization, reconstruction of technological equipment with detail, which is absent in the calculation of other indicators of energy efficiency.

2 Literature Review and Problem Statement

Various aspects of ensuring the efficient functioning of energy and its environmental impact are discussed in many papers [10–19]. Problems of efficient functioning and development of combined production of heat and electricity were addressed by famous scientists Sokolov Ye.Ya., Yakovlev B.V., Dolinsky A.O., Basok B.I.,

Klimenko V.N., Arkelyan E.K., Kozhevnikov N.N., Zaitsev Ye.D., Dubovsky S.V. and others [1, 20–29].

Existing methods for the distribution of energy consumption in multi-product industries have been described in many literary sources [1, 24, 25, 30], but the fuel distribution for CHP is considered only for the fuel burned in the boiler plant, without taking into account the energy costs for its storage, preparation, distribution. etc.

DSTU 3682–98 (GOST 30583–98) was introduced in Ukraine [9], which approved the indicator of total energy intensity of products. The Russian standards [31–33] introduce two indicators of energy intensity: total energy intensity of production and energy intensity of production. According to these standards, the energy intensity of production is the amount of energy and fuel consumption for the main and auxiliary technological processes of production based on a given technological scheme. When manufacturing any kind of products for each technological process that is part of the production schemes, its energy intensity is determined. Energy intensity in production of products is the integration of costs at a certain level of management. When calculating indicators of energy intensity in production, only fuel and energy resources (FER) for main and auxiliary production processes are taken into account. Expenditures on fuel and energy resources for heating, illumination, various auxiliary needs are not to be included in the amount of consumption of energy resources, which are included in the energy intensity of production. The Russian standard [33] classifies energy efficiency indicators by groups of homogeneous products (electric motors, turbines, refrigerators, etc.), type of energy resources or energy carriers used (energy efficiency indicators of fuel use (boiler, motor), electricity, thermal energy (steam, hot water, refrigerants), compressed air, oxygen, and other methods of indicators determination.

Full energy intensity determines the final energy saving, depends on the improvement of all components of the technological chain of production, development of new technologies, changes in the structure of production processes, reduction of material capacity and energy losses, increased use of secondary material resources, etc.

According to the standard of full energy intensity of products [9], it is defined as full energy consumption throughout the entire production chain: production of raw materials, transportation costs to the enterprise and for intra–plant transport, energy consumption for the main technological processes in which, in addition to direct costs of energy resources, energy resources produced in auxiliary shops are used (compressed air, oxygen, nitrogen, argon, circulating water supply for technological needs, steam and electricity production in industrial CHP plants, boiler houses and utilization plants), energy consumption, which is fixed in fixed production assets (equipment, appliances, premises, etc.), energy consumption of recreated workforce and on elimination of harmful influence on environment of production wastes.

In accordance with [9], the total energy intensity of products is calculated by the formula

$$e = e_e + e_{sm} + e_f + e_r + e_{env} \tag{1}$$

where e_e—full energy intensity of energy resources that directly are consumed for the production of products (services); e_{sm}—full energy intensity of original products, raw materials and materials required for the production of products (works, services); e_f—full energy intensity of fixed production assets amortized in the production of products (services); e_r—full energy intensity of recreated workforce in the production of products (services); e_{env}—full energy intensity of costs used for environmental protection in the production of products (services).

According to [9], direct energy intensity of products (full energy intensity of energy resources, through technological energy intensity) is defined in general terms as follows:

$$e_p = \sum_s e_s (b_{ps} + \sum_i a'_{pi} b'_{pis}), \qquad (2)$$

where s is the index of the type of energy resources; e_s—full energy intensity of s-th type of energy resources; b_{ps}—specific consumption of s-th type of energy resources in the main production, i—index of the type of auxiliary production; a'_{pi}—specific consumption of i-th type of auxiliary production; b'_{pis}—specific consumption of s-th type of energy resources for the production of i-th type of auxiliary production.

Full energy intensity of energy resources is defined as follows:

$$e_s = e_p + e_{tr} - e_g + e_{imp} \qquad (3)$$

where e_p—full energy intensity of energy resources directly consumed in the manufacture of products (services)—direct energy intensity; e_{tr} full energy intensity of energy resources used to transport the original products, raw materials and supplies; e_g—reduction of full energy intensity of products (services) through the use of secondary energy resources formed in the manufacture of combustible; e_{imp}—increase in full energy intensity due to the import of energy resources.

In fact, it is not known how e_s is defined in standard [9] since the algorithm for calculating the components of formulas (2–3) is not provided. For an example of calculating with formula (2), the energy intensity of energy carriers (electric and thermal energy as products) at CHP plant, b_{ps} is specific fuel consumption for the output of heat and electric energy separately. The second component (2) is electric power consumption for water feeding into steam generator and air feeding into furnace. As a rule, these specific costs are not calculated but taken as averages across the country from the statistical reporting forms. Type of fuel preparation, determined by the type of mill used for grinding solid fuel, type of fuel feeding system for solid fuel, liquid fuel heating, expenses for gas distribution networks as components of energy costs in the basic technological process of electricity and heat production in the formula are absent. If changes occur in these processes when replacing this equipment, then the reduction in unit costs cannot be calculated.

When calculating direct energy intensity of products (energy carriers) according to [9] it is not clear how to compare fuel combustion efficiency using different technologies (different burners and furnaces). The average rate of specific fuel consumption for heat and electricity output may not account for this. Made in boiler unit steam can be directed to different types of turbines (with different types of steam exhaust): industrial, heated, regenerative. Foreign CHP plants already use gas turbine units and steam-gas units (gas turbine unit of turbine waste heat recovery boiler). It also does not stand out when using the average specific fuel consumption for electricity and heat output. Besides, energy resources are also used for neutralization of emissions, waste and effluents. In other words, it is very difficult to calculate the change in energy consumption of energy carriers (electric and thermal energy) during modernization or detailed renewal of energy equipment according to the existing methodology. The existing methodology allows comparing technologies where certain components are different: technology for extraction (collection) of fuel (except solar and wind), type of fuel transportation (except solar and wind), type of energy generation (thermal, nuclear, solar, wind, biofuel). Energy costs for the production of electricity and heat itself are included in the component of full and direct energy intensity through the averaged across the country indicator of specific fuel consumption for the output of energy carriers without analysis of options for improving this indicator. Own requirements of CHP plants are not taken into account at all (electric and thermal energy). Among auxiliary expenses electric energy for water and air supply is taken into account.

A mathematical model for determining the full energy costs of production in natural, conditional or cost terms per unit of output (by the example of ferrous metallurgy), which clarified the concept of direct energy consumption of products was developed in the Institute of General Energy of the National Academy of Sciences of Ukraine [34]:

$$E = \frac{B_{o.p} - \left(B_{g.er} + B_{t.er} + B_{e.er}\right)}{N_{o.p}} - \frac{\sum_l B_{m.rl}}{N_{m.r}}$$

$$+ m_{c.m} \cdot \frac{\sum_j B_{c.mj}}{N_{c.m}} = \frac{\sum_s B_{ens}}{N_{en}} \tag{4}$$

where $B_{o.p}$—the volume of fuel and energy resources used for the manufacturing of products in the main and auxiliary production during the year, taking into account losses of energy resources in the technological process and the costs of intra-factory transportation (ton coal equivalent per year (tce/year)); $B_{g.er}$—the volume of combustible recycled energy resources received during the production of products (tce/year); $B_{t.er}$—amount of thermal recycled energy resources utilized during the year in the production of products (tce/year); $B_{e.er}$—amount of fuel, which was replaced by recycled energy resources of excess pressure in the generation of electricity produced during the year by utilization units (tce/year); $N_{o.p}$—amount of products produced by the enterprise during the year, in natural, conditional or cost terms; l—index of the type of recycled material resources, taken from the data of the enterprise; $B_{m.ri}$—the amount of energy resources, which accounts for the formation of recycled material resources in the technological process (tce/year); $N_{m.r}$—the amount of recycled material resources in physical, conditional or value terms, received during the year in the production of products; $m_{c.m}$—the share of raw materials, materials, intermediate products used in the technological process itself; j—index of type of raw materials, materials, intermediate products or assembly units; $B_{c.mj}$—amount of energy resources used during the year for the manufacture of j-ro type of raw materials, materials, intermediate products or assembly units, (tce/year); $N_{c.m}$—amount of j-type of raw materials, materials, intermediate products or assembly units in natural, relative or value terms, used during the year according to the data of the enterprise; m_{en}—amount of energy resource used directly in the technological process according to the data of the enterprise; s—energy resource type index; B_{ens}—amount of fuel and energy resources used for the production of s-type of energy resource (tce/year); N_{en}—amount of s-type of energy resource used for the year in natural, relative or value terms according to the data of the enterprise.

In multi-product industries for formulas (2–4) the distribution of energy expenses on a range of manufactured products in one technological unit should be performed. The most striking example, which requires the distribution of these costs, is primary refining. In [35] an analysis of existing methods for assessment of the energy efficiency of oil refining processes has been carried out on the basis of enlarged balances (energy, thermal and exergy) of the main technological units using various methods of distribution of total energy costs, and the exergy of fractions of primary oil refining at the output of the unit has been calculated. A methodological approach to determining the total energy consumption of oil products is provided, taking into account the technological scheme of the plant and the range of produced oil products. The algorithm for calculating the full energy intensity of oil products was defined as follows.

$$e_f^{op} = \sum_f \frac{m_{fr,f}^{op}}{m_f^{op}} \left[\frac{m_{fr,f}^{per}}{m_{oil}} \left(e_{oil}^{out} + e_{log}^{or} + e^{per} \right) + e_f^{sec} + \sum_g e_g^d \frac{m_{op,g}^d}{m_{raw}^d} \right]$$

$$+ \frac{m_f^{op}}{m_{oil}} \left(e_f + e_{env} \right) \tag{5}$$

where $m_{fr,f}^{op} \big/ m_f^{op}$—the portion of the intermediate product supplied to the blending unit of commercial oil products from the processes of primary, secondary and deep processing; $m_{fr,f}^{per} \big/ m_{oil}$—the portion of f-th fraction, comes out of the process of primary oil processing; e_{oil}^{out}—the direct energy intensity of the extracted oil that comes to the oil refinery for processing; e_{log}^{or}—the energy intensity of transporting oil to refineries; e^{per}—the full energy intensity of the primary oil processing process; e_f^{sec}—full energy intensity of the secondary refining process for f oil products in the technological chain of production; e_g^d—full energy intensity of the g-th process of deep processing; $m_{op,g}^d$, m_{raw}^d—the mass output of oil products and the bulk of processed raw materials in the g-th process of deep refining, respectively;

$m_f^{op} \big/ m_{oil}$—the portion of oil products obtained from oil in the balance of oil and petroleum products; e_f, e_{env}—as in formula (2).

It is proven that in the processes of physical fracturing of oil the most acceptable method of distribution of energy expenditure is in proportion to the mass output of fractions, because the thermal abilities of fractions are similar and slightly affect the refinement of distribution coefficients.

A similar approach was applied to the allocation of energy expenses for coke-chemical products [36], which made possible to calculate the energy intensity of 4 chemical products manufactured as side products during coke gas purification.

In the given examples (oil refining, coke chemistry), the received kinds of production are comparable on physical and chemical properties: have weight, density, temperature, pressure, their characteristics for various products can be measured by corresponding devices. When considering combined energy generation, two dissimilar products with different properties are produced and cannot be measured with the same instruments.

DSTU [37] developed in 2014 provides a methodical approach to the allocation of total fuel consumption at thermal power plants depending on the exergy of steam, which performs work in steam turbines of different types. This thermodynamic method makes it possible to determine the energy consumption of electric and thermal energy from the coefficient of thermodynamic value of heat, which depends on the type of steam turbine cycle (efficiency of steam exhaust) or gas turbine, but it is not possible to analyze the energy consumption efficiency from improvement of burner devices and systems of fuel preparation and feeding, as well as to use the potential of recycled energy resources of natural gas excess pressure again.

3 Purpose and Objectives of the Study

The purpose of the study is to analyze and improve the existing methodical approach to determining the direct end-to-end energy consumption on the example of energy carriers—electric and thermal energy, simultaneously produced at a combined heat and power plant from a steam turbine unit during the combustion of coal and natural gas, and include both the direct use of fuel to obtain steam, and support costs of energy resources for the preparation and supply of fuel to the burners of steam generators (own needs of the plant), reduction of which directly affects efficiency of energy production, use of recycled energy resources (RER), consideration of energy efficiency costs for environmental protection measures; use of improved approach to assessment of technological potential of energy saving in selection of efficient technologies.

The objective of the study is to analyze methodological approaches to determining the efficiency of energy costs in combined production, the use of indicators of technological energy efficiency along the production chain for energy saving potentials in various options for replacing the existing equipment with more efficient ones.

4   Improved Methodological Approach to Determination of Direct Technological Energy Intensity of Products

For multi-product industries, formula (2) takes the form:

$$e_p = k_n'' \sum_s e_s (b_{ps} + \sum_i a_{pi}' b_{pis}') + e_{env}^n, \qquad (6)$$

where k_n'' is the partition coefficient choice for a n-th specific multi-product manufacturing technology.

In the study is considered as combined energy production simultaneous production of thermal and electric energy. When splitting energy consumption into 2 energy products, different approaches can be applied, for example, the current one (50:50) and the approach that takes into account the seasonal demand for energy carriers, that is, the main product in the heating period will be thermal energy, by-product— electric. In the non-heating period, when the CHP will be included in the regulation of schedules of electric loads, the main product will be electricity, the by-product— thermal in the form of hot water or cold.

On the example of a coal-fired CHP plant with a T-type turbine with heating steam extraction for the needs of utility consumers, a chain of energy resources consumption was created from the supply of fuel to the CHP plant to the production of heat and electricity.

We have detailed the methodical approach for taking into account all the listed energy costs that can affect the energy intensity of energy carriers.

This structural sequence consists of the following steps [38]:

1. Determination of the main type of fuel (coal, fuel oil, natural or other industrial gases);
2. Definition of basic equipment;
3. Definition of auxiliary equipment;
4. Investigation of the efficiency of existing equipment and research of the possibility of introduction of the latest technologies;
5. Estimation of possible reduction of energy consumption when replacing existing technologies with promising ones.

An improvement in formula (2) is as follows. Not used as an indicator of energy intensity of energy resources specific fuel consumption for energy carriers supply of country. Energy intensity of energy resources is determined for a specific set of thermal equipment, taking into account the use of secondary thermal energy resources for coal CHP plants and the potential of secondary energy resources of excess pressure of natural gas for gas CHP plants [39]:

– coal CHP plant:

$$b_{sk}^{coal} = \sum_s \sum_k b_{sk} = b_{heat}^{sk} + b_{mill}^{sk} + b_{transp}^{sk} + b_{comb}^{sk} - b_{regen}^{sk} + b_{env}^{sk} \qquad (7)$$

– gas CHP plant:

$$b_{sk}^{gas} = \sum_{s} \sum_{k} b_{psk} = b_{transp}^{sk} + b_{comb}^{sk} - b_{regen}^{sk} - k_w W + b_{env}^{sk} \qquad (8)$$

where k—is the type of equipment; s—kind of energy resource; b_{heat}^{sk}, b_{mill}^{sk} b_{transp}^{sk} b_{comb}^{sk} b_{regen}^{sk} b_{env}^{sk}—respectively specific consumption of fuel (energy resource) for heating the fuel (for coal and heating oil), grinding coal at the used or new type of mill, supply of fuel through the pipeline, consumption for combustion, fuel savings as a result of the use of regenerative installations (air heater and water economizer) and energy costs to operate environmental equipment, depending on the type of burned fuel; $k_w W$—the amount of produced electricity in the turbo-expander due to the secondary energy resources of excess pressure (reduction of pressure at the gas distribution substation).

5 Analysis of Results Obtained

By formulas (6–7) the energy intensity of energy carriers (electric and thermal energy) is calculated together (Fig. 2) with allocation of all energy intensity components. Distribution of fuel consumption for generation of energy carriers should be done according to the method [37], while allocation of other fuel consumption can be done by different methods, e.g. proportional to the volumes of energy supply translated

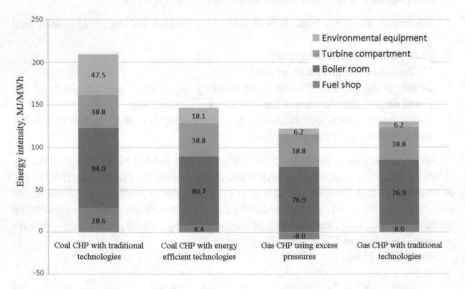

Fig. 2 The energy intensity of auxiliary costs of energy carriers (electricity and thermal energy) produced at the CHP plant depending on the type of equipment and used fuel

into the same units of measurement. It should also be added that the installed capacity utilization factor is applied to each type of equipment and at each stage, because not all equipment at the power plant is constantly in operation, but some equipment is operating only in a certain season (for example, defrosting of coal fuel in winter).

An information database was created for the implementation of this approach and model calculations, which contains a list of main and auxiliary equipment in various configurations, various manufacturers and with various energy indicators. By means of the given algorithm depending on type of the applied equipment energy intensity indicators, which can be compared and chosen more effectively as an indicator of decrease in energy intensity of energy carriers [38], are received.

Figure 2 shows the obtained calculations for production of 1 ton of steam by boiler units taking into account energy costs for fuel preparation and storage, operation of auxiliary equipment of boiler and turbine departments, as well as energy costs for environmental protection measures. The energy intensity of the burned fuel has not been presented as it remains unchanged when the auxiliary costs is reduced or increased. The energy intensity of the combusted fuel is 2718.38 MJ/t of produced steam or 4264.796 MJ/MWh of energy supply. An analysis of the possible replacement of auxiliary costs (without combustion fuel in the boiler) is included in the calculation model for the four variants of the CHP plants.

The first variant characterizes the technologies of a coal CHP plant with the most widespread main and auxiliary equipment currently operating in Ukrainian CHP plants using gas turbines for defrosting fuel in winter, ball mills for grinding, coal dusting with normal dust concentration in the air stream for flaring coals in chamber furnaces with hydro ash and slag removal, wet precipitation for particle removal, semi-dry lime technology for sulfur removal and selective catalytic reduction for reducing nitrogen oxide emissions.

The second case includes the improved coal power units with technologies which are partially used at separate thermal power plants and are perspective now, including the following main and auxiliary equipment have been put in place: gas radiation defrosting panels, roller or hammer mills, coal dusting with a high concentration of dust in the air stream for boilers with the technology of circulating fluidized bed allowed to organize measures on reduction of nitrogen oxides at once in the furnace at combustion without additional equipment.

From the data in Fig. 2 we can see that fuel preparation technology can significantly affect the overall energy intensity of products. It should also be noted that the combustion technology is more efficient, which allows to burn lower quality fuel with lower auxiliary energy expenses, which can be clearly seen when comparing the first and second variants.

The third option is a natural gas-fired combined heat and power plant using a turbo-expander at a gas distribution substation (GDU) while reducing pressure to a working one and the most common main and auxiliary equipment. Option 3 shows the full technological consumption of energy resources through the technological chain when using the energy of natural gas overpressure that emitted when gas pressure decreased to the worker pressure. According case Option 4, with which Option 3 is compared, the energy costs are higher by this value.

Energy intensity of environmental measures in direct energy intensity of energy resources is calculated according to the method [40] using the data from [41].

The results presented in Fig. 2 allow to determine energy saving potential (maximum possible fuel and energy saving) for thermal power plants at alternative replacement of auxiliary costs. For gas CHP plants, this potential is determined by additional energy, which can be produced by using recycled excess pressure energy resources.

For coal-fired cogeneration and power plants to estimate technological potential of energy conservation, it is possible to consider additional technology using an innovative turbine [42] (Table 1). This potential is determined by the product of the difference in energy intensity by variants of implementation volume (energy output). The difference in energy intensity is determined between the first and second and first and third variants. The energy saving potential can be determined for 1 MWh of energy carriers production (electric or electric and thermal energy—together).

The third variant is coal-fired power plants with the equipment listed in variant 2 (Fig. 2), in combination with a 100 MW electric turbine, which has a detachable coupling between medium and low pressure cylinders(CMP and CLP respectively), which has a higher efficiency (43.6% compared to 42–42.4%), can provide the power unit operation at sliding pressure at a reduced load, provides work on the thermal schedule 130/70°C without using peak water boilers. The use of a turbine with a split coupling allows increasing the maneuverability of the power plant, switching off the CLP during the heating period and eliminating ventilation losses in the CLP.

Table 1 Energy intensity of energy carriers before and after modernization of coal CHP plant, MJ/MWh

Equipment set options	Workshops of CHP plant				Energy intensity (technological) of energy carriers is an option
	Fuel shop	Boiler room	Turbine compartment	Environmental equipment	
1—Traditional technologies	28.59	93.97	38.85	47.48	208.89
2—Circulating fluidized bed furnace (CSF) and efficient assistive technologies	8.36	80.66	38.85	18.12	145.99
3—CSC furnace, detachable turbine and state-of-the-art assistive technology	8.36	80.66	28.17	18.12	135.31

Table 2 Energy saving potential for coal CHP plants from auxiliary equipment modernization, MJ of energy consumed/1 MWh of produced energy carriers together

Options	Energy intensity of energy carriers production (technological) at auxiliary equipment, MJ/MWh	Energy saving potential at alternative replacement of auxiliary technologies concerning Option 1, MJ/MWh
1—Traditional technologies[a]	208.89	0
2—Circulating fluidized bed furnace (CSF) and efficient assistive technologies	145.98	62.91
3—CSC furnace, detachable turbine and state-of-the-art assistive technology	135.30	73.59

[a]without the amount of fuel burned in the boiler, as its size does not change and is given above in the text

Coefficient of fuel heat utilization in the power unit under the third variant makes 95.17% against 88–89% for power units with traditional turbine unit [43].

The results of the assessment are presented in the Table 2.

The Table 2 shows, the technically possible energy saving potential at alternative replacement of auxiliary technologies for the production of energy carriers at coal-fired power plants can be achieved by upgrading the equipment according to option 2—62.9 MJ per 1 MWh of produced energy carriers, according to option 3—73.6 MJ per 1 MWh of produced energy carriers. Coal-fired power plants have a significant reserve to increase energy efficiency in the production of energy carriers. With a shortage of natural gas and unstable supplies of energy resources from renewable sources, coal-fired power plants can take their place in the structure of generating capacities.

6 Conclusions

The existing methodological provisions for determining the total energy intensity of the products and its components as a through-flow specific energy consumption was analyzed and compared.

Methodical approach to determining the direct energy intensity of products for multi-product energy intensive industries has been improved. The calculation is given for cogeneration of energy carriers at CHP plants, which burns different fuels. The proposed detailing has made it possible to calculate the amount of energy savings in the transition from widespread technologies to high-efficiency modern technologies for coal-fired CHP plants using three variants of applied technologies, and in natural gas—for two, taking into account the use of secondary energy resources of excess pressure for additional electricity production for own need.

Direct energy intensity includes energy intensity of enviromental measures, as the neutralization of pollutants occurs at the place of pollution formation, and the coefficient of distribution of common energy consumption at the point of distribution of products is introduced into the calculation algorithm.

An analysis of the latest technologies that can be implemented at coal power plants has shown a significant reserve for improving energy efficiency in energy carriers production. In the face of natural gas shortages and unstable energy supplies from renewable sources, coal power plants can take their place in the structure of generating capacity.

References

1. Dubovsky, S.V.: Energy-economic analysis of interconnected systems of electricity and heat generation. Naukova dumka, Kiev p. 182 (2014)
2. Report on energy supply and use. Ukraine, The State Statistics Service of Ukraine. Kiev (2017). Access mode: www.ukrstat.gov.ua
3. Report on use and stocks of fuel. Ukraine, The State Statistics Service of Ukraine. Kiev (2017). Access mode: www.ukrstat.gov.ua
4. Plachkov, I.S.: Types of thermal power plants. Access mode: http://energetika.in.ua/en/books/book-1/part-1/section-1/1-2
5. Smihula, A.V., Sigal, I.I., Bondarenko, B.I., Semeniuk, N.I.: Technologies for reducing harmful emissions to the atmosphere by thermal power plants and boilers of large and medium capacity of Ukraine, p. 108. FOP Maslak, Kyiv (2019)
6. National Emissions Reduction Plan for Large Combustion Plants. Website of the Ministry of Energy and Coal of Ukraine. Access mode: http://mpe.kmu.gov.ua/minugol/control/publish/article?art_id=245255506
7. DSTU 3755–98. Nomenclature of energy efficiency indicators and procedure for their inclusion in regulatory documents. Kyiv (1998)
8. Vagin, G.Ya., Dudnikova, L.V., Zenyutich, E.A.: Energy saving in industrial technologies (2001)
9. DSTU 3682–98 (GOST 30583–98). Energy saving. The methodology for determining the total energy intensity of products, works and services. Kyiv (1998)
10. Lytvynchuk, V.A., Kaplin, M.I., Bolotnyi, N. P.: The method of design an optimal under-frequency load shedding scheme. In: IEEE 6th International Conference on Energy Smart Systems, pp. 14–17 (2019) https://doi.org/10.1109/ess.2019.8764241
11. Kulyk, M.M., Kyrylenko, O.V.: The state and prospects of hydroenergy of Ukraine. Techn. Electrodyn. **4**, 56–64 (2019). https://doi.org/10.15407/techned2019.04.056
12. Popov, O.O., Iatsyshyn, A.V., Kovach, V.O., Artemchuk, V.O., Kameneva, I.P., Taraduda, D.V., Sobyna, V.O., Sokolov, D.L., Dement, M.O., Yatsyshyn, T.M.: Risk assessment for the population of Kyiv, Ukraine as a result of atmospheric air pollution. J. Health Poll. **10**(25), 200303 (2020). https://doi.org/10.5696/2156-9614-10.25.200303
13. Bilan, T., Rezvik, I., Sakhno, O., But, O., Bogdanov, S.: Main approaches to cable aging management at nuclear power plants in Ukraine. Nucl. Radiat. Saf. **4**(84), 54–62 (2019). https://doi.org/10.32918/nrs.2019.4(84).07
14. Babak, V.P., Babak, S.V., Myslovych, M.V., Zaporozhets, A.O., Zvaritch, V.M.: Technical provision of diagnostic systems. In: Diagnostic Systems For Energy Equipments. Studies in Systems, Decision and Control, vol. 281, pp. 91–133. Springer, Cham (2020). https://doi.org/10.1007/978-3-030-44443-3_4

15. Xin-gang, Z., Pei-ling, L.: Is the energy efficiency improvement conducive to the saving of residential electricity consumption in china? J. Clean. Produc, 249 (2020). https://doi.org/10.1016/j.jclepro.2019.119339
16. Mokhtar, A., Nasooti, M.: A decision support tool for cement industry to select energy efficiency measures. Energy Strat. Rev., 28 (2020). https://doi.org/10.1016/j.esr.2020.100458
17. Chakravarty, K., Kumar, S.: Increase in energy efficiency of a steel billet reheating furnace by heat balance study and process improvement. Energy Rep. **6**, 343–349 (2020). https://doi.org/10.1016/j.egyr.2020.01.014
18. Reuter, M., Patel, M.K., Eichhammer, W., Lapillonne, B., Pollier, K.: A comprehensive indicator set for measuring multiple benefits of energy efficiency. Energy Policy, 139 (2020). https://doi.org/10.1016/j.enpol.2020.111284
19. Leoni, P., Geyer, R., Schmidt, R.: Developing innovative business models for reducing return temperatures in district heating systems: approach and first results. Energy, 195 (2020). https://doi.org/10.1016/j.energy.2020.116963
20. Sokolov, E.Ya.: District heating and heat networks: a textbook for universities, 5th ed., Revised, p. 360. Energoizdat, Moscow (1982)
21. Yakovlev, B.: Improving the efficiency of district heating and heating systems, p. 448. Education and upbringing, Minsk (2002)
22. Utilities of Ukraine: state, problems, ways of modernization. In: Dolinsky, A.A., Basok, B.I., Bazeev, E.T., Pyrozhenko, I.A. (eds.), vol. 2, p. 820. Kyiv (2007)
23. Klimenko, V.N., Mazur, A.I., Sabashuk, P.P.: Cogeneration systems with heat engines: reference manual. In 3 parts. Part 1. General issues of cogeneration technologies, p. 560. Kiev, CPI ALKON NAS of Ukraine (2008)
24. Arakelyan, E.K., Kozhevnikov, N.N., Kuznetsov, A.M.: Tariffs for electricity and heat from the CHP. Heat Pow. Eng. **11**, 60–64 (2006)
25. Zaitsev, E.D.: Thermodynamic method for calculating the specific fuel consumption for various types of energy released by CHP. Heat Supp. News **12**(148), 24–26 (2012)
26. Dubovsky, S.V.: Increasing the maneuverability of the energy system through the introduction of heat pumps-regulators in the TPP. Prob. Gen. Energy **4**, 16–23 (2013)
27. Shubenko, O.L.: The transfer of a small cogeneration plant to burning local fuel in volumes that ensure its operation in the summer. Energy Sav. Encrgy Aud. **4**, 17–26 (2014)
28. Maliarenko, V.A.: Cogeneration technologies in the energy sector based on the use of low-power steam turbines: monograph. Institute of Engineering Problems of the NAS of Ukraine, Kharkov (2014)
29. Kesova, L.O., Horskyi, V.V.: Improving the efficiency of thermal power plants using low-cost technologies. Prob. Gen. Energy **2**(53), 60–64 (2018)
30. Brodyansky V.M.: Exergy method of thermodynamic analysis. Moscow, Energy (1973)
31. GOST R 50-605-100-94. Standardization Recommendations. Energy saving. The main directions of energy conservation in the steel industry. Technological measures to reduce the consumption of boiler and furnace fuel
32. GOST 51387–99. Energy saving. Normative and methodological support. The main provisions
33. GOST R 51541–99. Energy saving. Energy efficiency. General Provisions (2002)
34. Gnidoy, M.V., Kuts, G.O., Tereshchuk, D.A.: The method of calculating the total energy costs of production. Ecotechnol. Res. Conser. **5**, 67–72 (1997)
35. Gnidoy, M.V., Maliarenko, O.Ye.: Energy efficiency and determination of energy saving potential in oil refining. Kiev, Naukova Dumka (2008)
36. Maliarenko, O. Ye., Teslenko, O.I.: The use of the method of full energy intensity of products for the analysis of energy production efficiency. Prob. Gen. Energy **3**(23), 19–24 (2010)
37. DSTU 7674:2014 Energy Saving. The energy intensity of the technological process of generating electrical and thermal energy released by a thermal power plant. Method of determination. Kyiv (2014). 34 c
38. Horskyi, V.V., Maliarenko, O. Ye.: Methodical approach to the evaluation of the efficiency of modernization of the TPP of Ukraine. Access mode: http://molodyvcheny.in.ua/files/conf/other/37june2019/37june2019.pdf

39. Maliarenko, O. Ye., Horskyi, V.V.: An improved approach to the evaluation of the efficiency of energy saving measures and technologies at the TPP. Prob. Gen. Energy **4**, 24–31 (2019)
40. Stanytsina, V.V.: Development of full energy consumption method for determination of energy efficiency indicators and energy saving potentials. PhD thesis. Institute of General Energy of NAS of Ukraine, Kyiv (2016)
41. Artemchuk, V.O., et al.: Theoretical and applied bases of economic, ecological and technological functioning of energy objects (2017)
42. Horskyi, V.V., Maliarenko, O. Ye.: Estimation of energy saving potential for coal-fired thermal power plants in the implementation of innovative technologies. Research Practice Conference "Science, Technology and Technology: Global and Modern Trends." Prague, Czech Republic (2019)
43. New generation coal power plant. Access mode: https://docplayer.ru/26842769-Razrabotka-vysokoeffektivnyh-i-ekologicheski-chistyh-ugolnyh-tec-novogo-pokoleniya.html

Information-Measuring Technologies in the Metrological Support of Thermal Conductivity Determination by Heat Flow Meter Apparatus

Oleg Dekusha⊙, Zinaida Burova⊙, Svitlana Kovtun⊙, Hanna Dekusha⊙, and Serhii Ivanov⊙

Abstract The main efficiency indicators of heat-insulating materials and products are characteristics: thermal resistance and effective thermal conductivity. One of the main method of determination of thermal resistance described in EN 12667:2001 heat flow meter method. Important problem in using heat flow meter apparatus and method for measuring thermal conductivity is its calibration and testing. Usually this performed by using standard samples with known thermal conductivity but this gives only main information about state of apparatus. Proposed Information-measurement metrological unit which gives ability to perform recalibration of the apparatus without disassembling. The individual static conversion function of each of two heat flux sensors can be found by 'two measurement method'. The essence of the method is that the conversion function of the calibrated heat flux sensors is determined in two stationary thermal modes by measuring the power of energy supplied to a special calibration heat source placed between two heat flux sensors and the output signal of each sensor. Individual conversion characteristics of primary temperature sensors are determined during calibration by using information-measurement metrological unit gives possibility to reduce the contribution to the error value of measuring temperature. Proposed unit was tested in the information measurement system for determination of the thermal conductivity.

Keywords Information measuring system · Low conductivity insulators · Thermal resistance · Effective thermal conductivity · Temperature and heat flux sensors · Calibration

O. Dekusha (✉) · S. Kovtun · H. Dekusha · S. Ivanov
Monitoring and Optimization of Thermal Processes Department, Institute
Of Engineering Thermophysics NAS of Ukraine, 2a, Marii Kapnist Street, 03057 Kyiv, Ukraine
e-mail: ODekusha@nas.gov.ua

Z. Burova
National University of Life and Environmental Sciences of Ukraine, 15, Heroyiv Oborony Street, 03041 Kyiv, Ukraine

V. Babak et al. (eds.), *Systems, Decision and Control in Energy I*, Studies in Systems,
Decision and Control 298, https://doi.org/10.1007/978-3-030-48583-2_14

1   Introduction

The main efficiency indicators of heat-insulating materials and products are characteristics: thermal resistance and effective thermal conductivity.

There are many international normative documents regulating methods for determination the thermal resistance and related properties of various low conductivity insulators [1–5].

The stationary plate method for studying the thermal properties of insulators, recommended in world practice, is implemented in two ways that differ in the method for determining the heat flux through the test sample: the guarded hot plate method and the thermometric method. These methods are applied in most devices that are currently used to study the thermal resistance and conductivity of heat-insulating building materials [6–14]. In all apparatus designs the test sample is placed between the heat sources—heater and cooler using the conductive method of supplying and removing heat under the condition of reliable thermal contact between the contacting surfaces. Using these elements two isothermal surfaces should be created between which the sample must be located. Along the lateral surface of the sample the heat transfer also can be organized, both purely conductive and complex.

The guarded hot plate method is absolute by the heat conductivity determinate technique. The electrically heated measuring zone of the device's heating plate (central) which is surrounded on all sides by protective heaters controlled by differential thermocouples to prevent lateral heat losses. A unidirectional heat flux passing through the sample with a uniform surface density is created similar to the heat flux passing through an endless plate which is bounded by two plane-parallel isothermal surfaces. The heat flux density through a test sample is determined by the electric power value, that is given on the devices central heater, its square is known, and the thermal conductivity can be calculated. Basic limitations during work on the guarded hot plate method devices consist in difficulties of continuous support the one-directed heat flux with a permanent density, exact measuring of power, that is given on a central heater, and temperatures, and also in a necessity to provide the maximal identity of structure, thickness and surface quality of both samples at the use of symmetric cell. In addition, large attention it follows to spare to quality of tangent surfaces contact planes for the contact thermal resistance minimization. The standards [1–3] indicate restrictions on the sample dimensions: the diameter or side of the square should be 0,3 or 0,5 m. When examining the thermal conductivity of homogeneous materials, this size can be 0,2 m, and in case of large thickness testing material 1 m. The limit of the supposed relative measurement error at room temperature on a guarded hot plate apparatus should be 2%, and in the entire temperature range no more than 5%.

The second approach heat flow meter apparatus, requires the use of additional special heat flux sensors (HFS) along with temperature sensors (TS). A heat flow created by heater and cooler is one-directed and pierces simultaneously the central zone of test sample and the sensitive zone of one or two identical heat flux meters.

This approach greatly simplifies the measurement process while reducing the number of necessary secondary devices and that is advantage in mass technical measurements. Another important advantage of devices equipped with HFS and used to measure thermal resistance and coefficient of effective thermal conductivity on large samples of heat-insulating building materials is a shorter duration of its operation until stationary conditions appear.

Important problem in using heat flow meter apparatus is its calibration and testing. Usually this performed by using standard samples [3] but this gives only main information about state of apparatus. In presented paper proposed metrological unit which give information about individual static conversion functions of the measuring heat flux and temperature sensors. Also this module gives ability to perform recalibration of the apparatus by absolute method.

2 Steady-State Heat Flow Meter Information Measurement System

Theoretical calculations and practical research [16–18] in the field of thermal resistance and conductivity of insulating materials studies made it possible to create a steady-state heat flow meter information measurement system. A symmetric scheme of the stationary method for determining the thermal resistance of low conductivity materials according to the requirements of standards [3–5] using primary heat flux sensors (HFS) [7, 15, 19] was implemented. The heat flow meter apparatus measurement cell structure is presented in Fig. 1.

A sample of the test material is located between two thermometric blocks. Each block is a plate containing primary temperature and heat flux sensors. The heat flux through the sample is set using an electric heater made of metal with high thermal conductivity. This contributes to isothermal conditions on the heat transfer surface of the heater in contact with the sample. The heat removal from the lower heat

Fig. 1 The heat flow meter apparatus cells structure

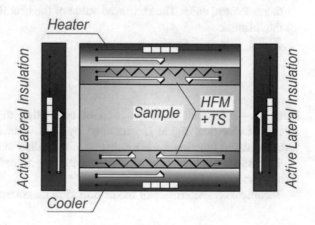

transfer surface of the sample is carried out by a cooler. Active lateral insulation, the temperature of which is maintained equal to half the temperature of the heater and cooler, creates adiabatic experimental conditions, eliminates the influence of external factors on the thermal field of the sample, and minimizes lateral heat loss.

Thermal resistance and thermal conductivity can be determined by the primary sensors signals measuring. The individual characteristics of these sensors should be experimentally established in calibration experiments.

As indicated above, for a symmetric measurement cell structure using two identical HFS and temperature sensors, the calculation formulas for thermal resistance R and effective thermal conductivity coefficient λ are:

$$R = \frac{\Delta T}{\bar{q}} - R_{SC} \tag{1}$$

$$\lambda = \frac{h}{R} \tag{2}$$

were $\Delta T = T_H - T_C$—temperature values difference between hot T_H and cold T_C sample working surfaces, K;

$\bar{q} = 0,5 \cdot (q_H + q_C)$—superficial heat flux density mean value through hot q_H and cold q_C sample working surfaces, W/m^2;

R_{SC}—summary contact heat resistance correction of measuring cell determined during device calibrating, m^2·K/W;

h—test samples thickness, m.

In formulas (1, 2), the quantities h, T_H, T_C, q_H, q_C are measurable directly. Their values can be measured by:

- the sample thickness by a device of a certain accuracy, for example, an electronic caliper,
- temperatures by primary thermoelectric temperature sensors—individually calibrated thermocouples type E,
- surface heat flux density by HFS mounted respectively in the upper and lower thermometric units. The measured value of the heat flux density is calculated by the formula:

$$q = K_q \cdot E \tag{3}$$

were K_q—individual static function of converting the heat flux density into a measured electrical signal E generated by HFS. Under the condition of a linear transformation function K_q is a constant value called the conversion coefficient.

The information measurement system for determination of the thermal conductivity is shown in Fig. 2 [17].

Technical characteristics of the information measurement system are:

Fig. 2 The information measurement system for determination of the thermal conductivity [16] 1—thermal unit, 2—electronic unit, 3—stabilisation thermostat

1 – thermal unit, 2 – electronic unit, 3 – stabilisation thermostat

- thermal conductivity coefficient measurements range is from 0,02 to 3,0 W/(m K);
- main relative error is ± 3%;
- operating temperature from range is from –40 to +180 °C;
- sample size is 300 × 300 × (10…120) mm.

3 Information- Measurement Metrological Unit and It Practical Application

For the purpose of calibrating a steady-state heat flow meter apparatus to measure thermal resistance and effective thermal conductivity proposed to create information-measurement unit based on the modular construction principle.

Modular construction principle gives possibility to work in different modes as well simplifies maintenance, extends the functionality, and improves the accuracy and reliability. First mode used to determine the individual static conversion functions of the measuring heat flux sensors. Second mode determine the individual conversion functions of the temperature sensors. Third mode is correction for contact thermal resistance.

The structure of information- measurement unit hardware module shown on Fig. 3.

For implementation of temperature and heat flux sensor signals recording, used modules with the following characteristics:

- 8-channel ADC with a bit of 16 bits and a conversion rate of 10 Hz;
- dynamic range setting and calibration;

Regualtor of heat source used DC voltage regulator.

Voltage signal recording was performed by Multimeter digital Picotest M3500A with relative error of measurement of voltage 0,0020/ 0,0006% in the range of 100 V and 1000 V.

Current signal recording was performed by Multimeter digital Picotest M3500A with relative error of measurement of current 0,05% in the range 3 A.

Fig. 3 The structure of
information- measurement
unit hardware

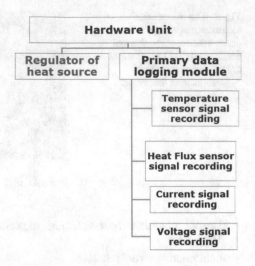

As additional control module used Multimeter digital Fluke 8846 A.

The software package implements the process of the interface management, data codes of analog signals processing and storing information.

3.1 Determination of the Individual Conversion Function of Heat Flux Sensors (HFS)

To determine the individual static conversion function of each of two identical HFS the 'two measurement method' was applied. The essence of the method is that the conversion function of the calibrated HFS is determined in two stationary thermal modes by measuring the power of energy supplied to a special calibration heat source placed between two HFS and the output signal of each sensor. The advantage of this method is the ability to calibrate thermometric devices containing several HFS in a wide range of operating temperature values passing it stepwise with any step.

To ensure uniform density of the heat flux passing through the surface of the calibrated HFS in contact with the heater, the calibration heat source—an electric heater—should be performed in compliance with certain requirements, namely: be anisotropic so that its thermal conductivity in the axial direction is several times greater than radially, and heat sources are distributed evenly on its working surfaces. Based on these requirements, a flat electric heater was made from a constantan wire 0.3 mm in diameter, laid on a specially made matrix with a pitch of 1 mm and filled with an insulating epoxy compound.

A schematic diagram of HFS calibration included in the stationary measuring system is shown in Fig. 4.

The method is implemented under conditions of conductive heat transfer in a thermostat located in the measuring cell (see Fig. 4), consisting of two thermometric

Fig. 4 Schematic diagram the HFS calibration by the 'two measurements method'

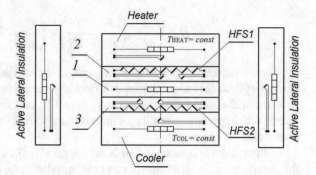

units 2 and 3 with a calibration heat source 1 between them. The areas of contacting surfaces of the heat source and both thermometric units are the same. The constant temperature difference between the heater and the cooler is set.

Two stationary thermal conditions are set sequentially by inputting two different power values to the calibration heat source. In each mode the heat flux passes through HFS1 and HFS2 and for its density values (q_1 and q_2—in the first mode, q_1' and q_2'—in the second) the expressions are valid based on formula (3):

$$q_1 = K_{q1} \cdot E_1, \ q_2 = K_{q2} \cdot E_2 \tag{4}$$

$$q_1' = K_{q1} \cdot E_1', \ q_2' = K_{q2} \cdot E_2' \tag{5}$$

where K_{q1}, K_{q2}—calibration factors, respectively, of the upper and lower HFS;

$E_{1 \text{ or } 2}$ и $E_{1 \text{ or } 2}'$—values of the output signal of the corresponding HFS in steady state in the first and second modes.

Thus, the following system of equations of thermal balance for the thermal column holds:

$$\begin{cases} P = K_{q1} \cdot E_1 \cdot F - K_{q2} \cdot E_2 \cdot F - \text{ for the first mode} \\ P' = K_{q1} \cdot E_1' \cdot F - K_{q2} \cdot E_2' \cdot F - \text{ for the second mode} \end{cases} \tag{6}$$

where P, P'—power of thermal energy generated by a heat source in two consecutive thermal modes calculated for an electric heater according to a known ratio $P = I \cdot U$;

F—the surface area of the zone of the calibration heat source in which the heating element is placed.

The solution of system (6) with respect to the calibration coefficient K_{q1} of upper HFS1 and K_{q2} lower HFS2 allows to determine their values according to the following formulas:

$$K_{q1} = \frac{1}{F} \cdot \left(\frac{P}{E_2} - \frac{P'}{E_2'} \right) \cdot \left(\frac{E_1}{E_2} - \frac{E_1'}{E_2'} \right)^{-1} \tag{7}$$

$$K_{q2} = \frac{1}{F} \cdot \left(\frac{P}{E_1} - \frac{P'}{E_1'} \right) \cdot \left(\frac{E_2'}{E_1'} - \frac{E_2}{E_1} \right)^{-1} \tag{8}$$

When calibrating devices at each point in the temperature range, the automatic control system sets and maintains a constant temperature difference between the heater and the refrigerator, equal to 6 … 10 K, and the temperature value of the lateral active thermal insulation, which is equal to half the temperature values of the heater and the cooler.

Further in the first mode power is not supplied to the electric heater, i.e. $P = 0$. In the second mode, such a value of power is supplied to the calibration electric heater to ensure that more than 95% of the heat flux flows through HFS2. Calculation formulas (7) (8) are converted to:

$$K_{q1} = \frac{P'}{F} \cdot \left(E_1' - E_1^0 \cdot \frac{E_2'}{E_2^0} \right)^{-1} \tag{9}$$

$$K_{q2} = \frac{P'}{F} \cdot \left(E_2^0 \cdot \frac{E_1'}{E_1^0} - E_2' \right)^{-1} \tag{10}$$

where E_1^0, E_2^0—signals of the corresponding HFS in the "zero" mode.

Based on the results of theoretical calculations a constant-nickel galvanic pair HFS were developed. Despite the unconventional combination of these two thermoelectric materials having the thermal electromotive force of the same sign relative to platinum, HFS based on them have high temperature and temporary stability. The calibration characteristics of such HFS obtained in the experimental series of measurements in the operating temperature range of -25 … $140\,°C$, described by polynomials, are presented in Fig. 5.

The graphs in Fig. 5 confirm the high temperature stability of constantan-nickel HFS. Their maximum discrepancy between the values of the conversion coefficients is no more 8 … 10% over the entire range of operating temperature.

3.2 Definition the Thermocouples' Individual Conversion Functions

The standard static characteristic of primary temperature sensors type E, normalized in [20], has a limit of permissible deviations of thermo-EMF from NSC $\pm 2.5\,°C$, which in the range of operating temperature values of the device will make a significant contribution to the error value. Therefore, despite the fact that TS measuring instruments are made of standard thermoelectrode wire, for each of them, individual conversion characteristics are determined during calibration of the measuring system. In addition, to increase the accuracy of temperature measurement, all TS are made differential, and their reference junctions are thermostated

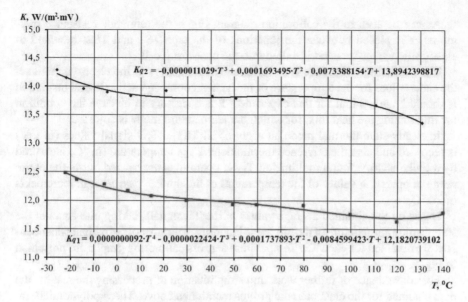

Fig. 5 Conversion functions of constant-nickel HFS represented by the polynomials

The temperature control and regulation system consists of the following sensors divided into four characteristic groups:

upper and lower elastic gasket TS1 and TS2,

upper thermometric unit measuring TS3 and control TS4,

lower thermometric unit measuring TS5 and control TS6,

three control TS7, TS8 and TS9 located in the heater, cooler and lateral insulation respectively.

These sensors are located as shown in the calibration diagram Fig. 6.

An individual TS calibration experiment is performed applying a reference temperature sensor, for example, a Pt100 platinum resistance sensor, the working standard in the temperature range from minus 50 to 156 °C, mounted in a thin plate for convenient placement in the measuring system thermal block measuring cell.

Fig. 6 Calibration scheme
of temperature measuring
sensors

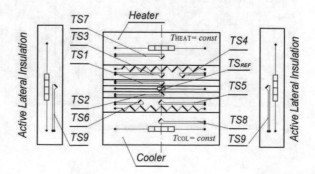

As can be seen in the calibration diagram Fig. 6 the reference resistance thermometer is placed between the junctions of the tape TS1 and TS2, mounted in elastic silicone gaskets and installed into the measuring cell.

The calibration procedure is as follows. The regulators of the electronic unit set the same values for the temperature of the heater, cooler and active lateral insulation It should be noted that for this experiment it is necessary to observe the condition for minimizing the heat flux, for which the temperature mode is adjusted.

In steady-state thermal mode an array of all TS1... TS9 signal values (in mV) is recorded and also the reference thermometer T_{REF} temperature (in °C) measured by a Fluke 8846A digital multimeter. These operations are carried out in the entire range of operating values of the temperature of the device, passing it in increments of 10 °C.

Based on the obtained average values of the TS signals, taking into account the stabilization temperature of the reference junctions, a number of individual dependencies are obtained for each TS. Temperature sensors calibrating graphs are presented in Fig. 7.

For convenience of further work and simplification of processing the results, the data obtained for the characteristic groups mentioned above. These polynomials are inputted into a software for measurement data processing.

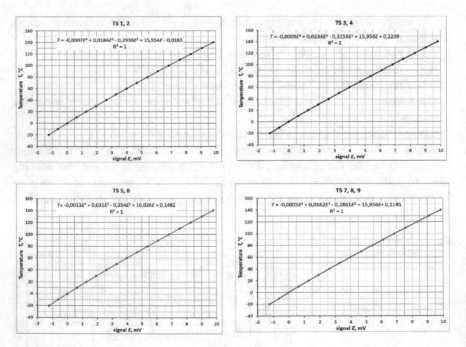

Fig. 7 Temperature sensors calibrating graphs

3.3 Determination the Total Contact Thermal Resistance Correction

When calculating the thermal conductivity coefficient of the test sample, the calculation formula (1) includes a correction for the total contact thermal resistance R_{SC} due to two main factors:

- the presence of microroughnesses on both contact planes of the test sample with the upper and lower thermometric blocks;
- design features of thermometric units, in which measuring temperature sensors are mounted under a thin layer of molding compound.

To minimize the influence of surface roughness of solid samples, which are studied in stationary devices under normalized load a set of elastic gaskets with primary temperature sensors is used. Also in the study of homogeneous solid samples, a highly thermally conductive lubricant, for example, silicon-organic oil is applied to the junction of each TS at the contact point. All of the above makes it possible to reduce the value of the total contact thermal resistance to about 10^{-5} (K·m^2)/W, which can be neglected in the calculations of the thermal conductivity coefficient.

The soft and fibrous materials samples effective thermal conductivity study is carried out according to the recommendations of [3–5] at low load or its absence. The samples structure themselves does not allow of any lubricants using, so the use of silicone gaskets in this case is useless. Such samples are placed in the measuring cell of the thermal block of the measuring system directly between the thermometric blocks, the surface roughness of which for these samples does not affect the total contact thermal resistance. Therefore, it is necessary to determine the correction for the total thermal resistance of the layers of the electrical insulating casting compound, which are located between the surfaces of the upper and lower temperature sensors and junction of the corresponding measuring TS. It called 'ballast resistance' of the measuring cell.

To conduct an experiment to determine the ballast resistance of a measuring cell, the schematic diagram of which is shown in Fig. 8, a set of elastic gaskets was used,

Fig. 8 Schematic diagram for the ballast resistance determining

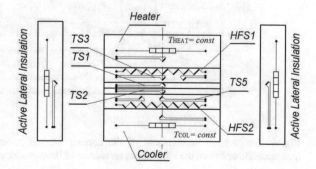

TS junctions of which in this study were placed on the surfaces of thermometric blocks, and the blocks themselves were brought together under load.

Then a series of measurements was carried out over the entire range of the operating temperature of the measuring system, passing it with a step of 10 °C. In each steady-state thermal mode an arrays of heat flux density values was recorded that passed through HFS1 and HFS2 the upper and lower thermometric units (q_1, q_2) and temperature values measured by the elastic gaskets TS1 and TS2 and the thermometric units TS3 and TS5 (Fig. 8).

According to the averaged experimental data the thermal resistance of the compound layers of the upper and lower thermometric blocks R_{C1} and R_{C2} was calculated using the formulas:

$$R_{C1} = (T_{TS1} - T_{TS3})/q_1 \qquad (11)$$

$$R_{C2} = (T_{TS2} - T_{TS5})/q_2 \qquad (12)$$

The thermal resistance value R_{SC} of the measuring cell is obtained by adding the counting results. The results of determining the thermal ballast resistance of the measuring cell obtained for a stationary device are graphically presented in Fig. 9.

As seen in Fig. 9, after processing the measurement information, the obtained function $R_{SC} = f(T)$ is described by the first degree polynomial. This polynomial is inputted into the computer program for calculating the thermal conductivity coefficient as a correction in the study of samples performed without the elastic gaskets using.

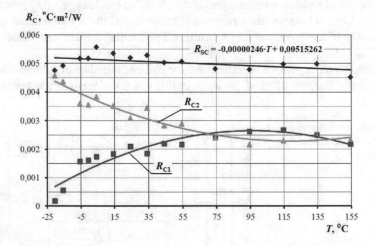

Fig. 9 Temperature dependences of the contact thermal resistance of the upper and lower thermometric units and the total ballast resistance of the measuring cell

4 Conclusions

Important problem in using heat flow meter apparatus for measuring thermal conductivity is its calibration and testing. Usually this performed by using standard samples with known thermal conductivity but this gives only main information about state of apparatus. Proposed Information-measurement metrological unit which gives ability to perform recalibration of the apparatus without disassembling.

The individual static conversion function of each of two heat flux sensors can be found by 'two measurement method'. The essence of the method is that the conversion function of the calibrated heat flux sensors is determined in two stationary thermal modes by measuring the power of energy supplied to a special calibration heat source placed between two heat flux sensors and the output signal of each sensor. The advantage of this method is the ability to calibrate thermometric devices containing several heat flux sensors in a wide range of operating temperature values passing it stepwise with any step.

The standard static characteristic of primary temperature sensors type E has a limit of permissible deviations of thermo-EMF from NSC $\pm$ 2.5 °C, which in the range of operating temperature values of the device will make a significant contribution to the error value. Individual conversion characteristics are determined during calibration by using information-measurement metrological unit gives possibility to reduce this contribution.

Proposed unit was tested in the information measurement system for determination of the thermal conductivity.

References

1. ISO 8302:1991, Thermal insulation—Determination of steady-state thermal resistance and related properties—Guarded hot plate apparatus
2. ASTM C177-13, Standard Test Method for Steady-State Heat Flux Measurements and Thermal Transmission Properties by Means of the Guarded-Hot-Plate Apparatus
3. EN 12667:2001, Thermal performance of building materials and products—Determination of thermal resistance by means of guarded hot plate and heat flow meter methods—Products of high and medium thermal resistance
4. ISO 8301:1991, Thermal insulation—Determination of steady-state thermal resistance and related properties—Heat flow meter apparatus
5. ASTM C518-10, Standard Test Method for Steady-State Thermal Transmission Properties by Means of the Heat Flow Meter Apparatus, Developed by Subcommittee: C16.30. Annual Book of ASTM Standards, vol. 04.06. https://doi.org/10.1520/c0518-10
6. Zarr, R.: A History of Testing Heat Insulators at the National Institute of Standards and Technology. ASHRAE Transactions, 107 (2001)
7. Akos, L., Ferenc, K.: Analysis of water sorption and thermal conductivity of expanded polystyrene insulation materials Building Serv. Eng. Res. Technol. 34(4), 407–416 The Chartered Institution of Building Services Engineers (2012) https://doi.org/10.1177/014362441246 2043bse.sagepub.com
8. Burova, Z.: Methods and devices for insulation materials heat-conducting determination. SWorld J. 4(11), 26–28 (2016)

9. Zarr, Robert, Carvajal, Sergio, Filliben, James: Sensitivity Analysis of Factors Affecting the Calibration of Heat-Flow-Meter Apparatus. J. Test. Eval. **47**, 20170588 (2019). https://doi.org/10.1520/JTE20170588
10. Line Heat-Source Guarded Hot Plate. https://www.nist.gov/laboratories/tools-instruments/line-heat-source-guarded-hot-plate,last, accessed: 2020/02/16
11. Thermal Conductivity Test Tool λ-Meter EP500e. http://www.lambda-messtechnik.de/en/thermal-conductivity-test-tool-ep500e.html, last accessed: 2020/02/16
12. GHP 456 Titan®. https://www.netzsch-thermal-analysis.com/en/products-solutions/thermal-diffusivity-conductivity/ghp-456-titan, last accessed: 2020/02/16
13. HFM 446 Lambda Series, https://www.netzsch-thermal-analysis.com/en/products-solutions/thermal-diffusivity-conductivity/hfm-446-lambda-series, last accessed: 2020/02/16
14. All New LaserComp FOX Series Heat Flow Meters. https://www.tainstruments.com/all-new-lasercomp-fox-series-heat-flow-meters, last accessed: 2020/02/16
15. Cesaratto, P.G., De Carli, M.: A measuring campaign of thermal conductance in situ and possible impacts on net energy demand in buildings. Energy Build. **59**, 29–36 (2013)
16. Burova, Z., Vorobyov, L., Dekusha, L., Dekusha, O.: Complex ИТ-7С for building materials thermal conductivity measurements. Metrol. Ins. Kharkiv **6**, 9–15 (2009). (in Ukrainian)
17. Babak, V., Burova, Z., Dekusha, O.: Calorimetric quality control of heat insulating materials. Int. J. NDT Days. Bulgarian Soc. NDT **1**(4), 469–476 (2018)
18. Hardware-software for monitoring the generation, transportation and consumption of thermal energy: Monograph, p. 298. Ed. prof. doct. V. Babak, NAS of Ukraine (2016). ISBN 978-966-02-7967-4. (in Ukrainian)
19. DSTU 3756–98,Energy saving. Heat flux sensors thermoelectric for general purpose. General specifications
20. IEC 60584–1:2013, Thermocouples—Part 1: EMF specifications and tolerances

Modeling of Power Systems with Wind, Solar Power Plants and Energy Storage

Mykhailo Kulyk and Oleksandr Zgurovets

Abstract This paper describes the process of frequency and power regulation in integrated power systems with wind, solar power plants and battery energy storage systems. A mathematical model consisting of the general power balance equation, equations describing the operation of all generators, electrical load and losses, as well as limitations and initial conditions has been developed for the research. The obtained model allows studying the features of frequency and power regulation during operation of wind, solar power plants and battery energy storage systems for different conditions and characteristics of system elements, as well as determining the most effective laws of regulation. Calculations were performed on an aggregated scheme of the power system, which includes nuclear and thermal power plants, hydroelectric power plants, loads, losses, wind, solar power plants and battery energy storage systems. It is shown that introduction of wind and solar power plants of large capacity into the structure of generating capacities of the integrated power system without taking additional measures on power balancing can lead to unacceptable frequency deviations in the system. It was proved that stabilization of frequency and power in integrated power systems with powerful wind and solar power plants can be achieved by introducing into the structure of integrated power systems of battery energy storage systems with a capacity comparable to the installed capacity of renewable energy sources. The results of modeling the operation modes of the mentioned power systems showed that the accuracy of frequency and power regulation in them is achieved higher than the standard for the EU power system ENTSO-E.

Keywords Mathematical model · Integrated power system · Wind power plant · Solar power plant · Battery energy storage system · Frequency and power regulation

M. Kulyk · O. Zgurovets (✉)
Institute of General Energy of NAS of Ukraine, Kyiv, Ukraine
e-mail: zgurovets_ov@nas.gov.ua

© The Editor(s) (if applicable) and The Author(s), under exclusive license
to Springer Nature Switzerland AG 2020
V. Babak et al. (eds.), *Systems, Decision and Control in Energy I*, Studies in Systems,
Decision and Control 298, https://doi.org/10.1007/978-3-030-48583-2_15

231

1   Introduction

The desire to reduce the negative impact on the environment associated with greenhouse gas emissions, harmful substances, anthropogenic load caused by extraction and subsequent production of primary energy resources, disposal of spent fuel, including nuclear, accidents at traditional energy facilities, as well as high prices and exhaustiveness of fossil energy resources push the world community to strengthen the development of renewable energy sources, increasing their role in the global energy balance.

Despite the sustainability, inexhaustibility and price reduction, along with comprehensive government and financial support, sources such as wind and solar power have a major drawback that still holds back the massive introduction of these systems around the world instead of traditional fossil fuel power plants. This disadvantage is the unevenness of electricity production, which is directly dependent on weather conditions: the intermittent (unstable) nature of the renewable energy source. The intensity of generation often does not coincide with peak hours, and the lack of inertia in photovoltaic power plants means that the generation power, which is directly proportional to the insolation of solar cells, can change rapidly over a wide range of tens of seconds several times a day.

The peculiarity of power systems operation is the simultaneous generation and consumption of electricity, which requires a constant balance between consumers and generators. Integration of wind (WPP) and solar (SPP) power plants (intermittent generation) into the energy system will require the installation of an appropriate amount of regulating capacity or changes in consumption patterns.

The main methods of generation management include the creation of power reserves and the introduction of maneuver units to maintain frequency, which is quite expensive in terms of increased fuel consumption and equipment wear. Approaches aimed at addressing the issue of uneven consumption include the stimulation of consumers, the introduction of restrictions, the creation of controllable loads [1, 2], which is not always effective in terms of frequency control, or difficult technically.

The integration of renewable sources also considers various measures to limit output power or shut down units in case of excess generation, which leads to a decrease in the efficiency of generation or a decrease in the utilization rate of equipment. The possibility of geographically dispersing and integrated use of such installation in conjunction with other more predictable generation sources as bioenergy, combined with more accurate generation levels planning with accurate weather forecasts, is also considered, which does not always have a positive effect or has limitation. Another rather promising solution to the problem of not only the integration of renewable sources, but also the traditional issue of frequency and power management in the grid, can be electricity accumulation, which until recently has been difficult to implement on an industrial scale.

Hydropower is the most common energy storage technology, but its use is often limited by the lack of hydro resources and environmental issues related to the construction of large reservoirs. Energy storage in the form of compressed gas, electrolysis hydrogen, large flywheels or accumulated heat has several disadvantages, which include insufficient study, high cost, or low efficiency, which is why they are still poorly distributed. The development of large-scale electricity storage technologies in the form rechargeable batteries in areas such as transport and energy, suggest to talk about the possibility of competing renewable wind and solar energy with traditional fossil fuel sources [3]. The battery energy storage systems (BESS) also provides high speed of changing output power, which sets them apart from other types of regulators in order to maintain the balance of generation and consumption in the power system and to stabilize frequency.

2 Literature Review and Problem Statement

Frequency regulation in power systems with intermittent power supply and energy storage is dedicated to many scientific works of such scientists: Anzalchi A., Arani M.F.M., Argyrou M.C., Arul P.G., Asghar F., Awais M., Bahloul M., Bisht M.S., Bolotnyi N.P., Brand U., Cho S., Christodoulides P., Dai Y., Dechanupapritta S., Duan X., El-Saadany E.F., Habib M., Han Y., Hao X., Harrag A., Hassan S.Z., Indu P.S., Jamroen C., Jang B., Jayan M.V., Jeon W., Kalogirou S.A., Kamal T., Kaplin M.I., Khadem S.K., Kim C., Kim S.H., Kircher K.J., Kulyk M.M., Kyrylenko O.V., Ladjici A.A., Li D., Li H., Li Y., Lytvynchuk V.A., Marouchos C.C., Maydell K.V., Melo S.P., Orihara D., Passerini S., Ponciroli R., Pour M. M., Qian F., Ramachandaramurthy V.K., Saitoh H., Sarwat A., Schuldt F., Shi D., Shi L., Spanias C., Talha M., Telle J.S., Vilim R.B., Vogt T., Wang J., Wang M., Wen Y., Xie P., Xu H., Yan S., Yang T., Yang X., Yoon Y., Zhang K.M., Zhang Z., Zhou T., Zhou X., Zhu L. and other [1–30].

Paper [5] proposes a combined virtual inertia emulator (VIE) and a hybrid battery-supercapacitor-based energy storage system (Fig. 1) for enhancing the stability of the Microgrids and smoothing the short-term power fluctuations simultaneously.

Authors [6] compare the advantages and disadvantages of using DFIG rotating mass or super-capacitor as the virtual inertia source and show that while virtual inertia is not incorporated directly in long-term frequency and power regulation, it may enhance the system steady-state behavior indirectly.

Model [7] contains a Building Integrated Photovoltaic (BIPV) system connected to the grid through a DC-DC boost converter, a DC-AC inverter and a battery storage system (Fig. 2) in active parallel configuration. The proposed model is implemented and verified through several simulations in Matlab/Simulink. Other approaches to the use of energy storage for frequency regulation in power systems are shown in articles [12, 14–18, 21–23, 26, 27].

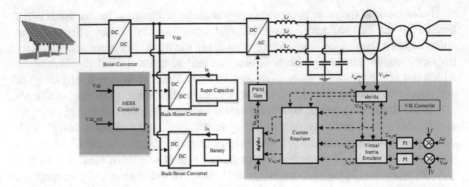

Fig. 1 An overview of the proposed system **a** Hybrid Energy Storage Controller **b** Virtual Energy Inertia Controller [5]

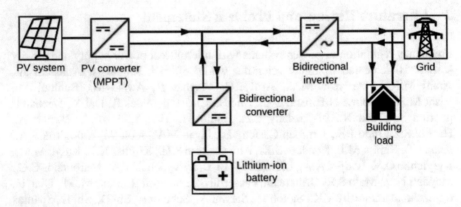

Fig. 2 Grid-connected PV system with battery energy storage [7]

In article [8] shows the potential challenges facing the operation of a standalone hybrid system with solar photovoltaic (PV) and wind turbine energy generation and battery energy storage (Fig. 3). Also developed control strategies for the system sources to mitigate the potential challenges.

In paper [9], a newly proposed fuzzy logic-based robust control mechanism is used to stabilize the frequency and direct current bus voltages in large fluctuations caused by sudden changes in power generation or load side. Also, supercapacitor and battery energy storage system are used to stabilize direct current bus voltages. Similar issues with fuzzy logic-based approaches and similar schemes (Fig. 4) are also addressed in the articles [11, 13, 15].

Very recently in the UK, the distribution network operator, National Grid (NG) have developed a novel portfolio for providing primary active power ancillary services known as Enhanced Frequency Response (EFR) service. In this paper, a design guideline of electrical energy storage system (ESS) for this EFR provision is given. In addition, a power management system is described to ensure the control

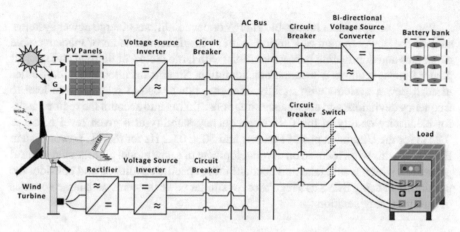

Fig. 3 Basic configuration of the standalone hybrid renewable energy system [8]

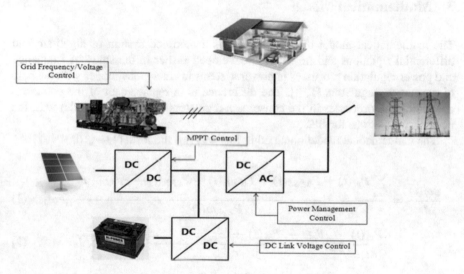

Fig. 4 Grid-connected hybrid power system [13]

of a hybrid energy battery-supercapacitor storage system (HESS). Simulations are carried out for both batteries only energy storage system (BESS) and HESS to analyze the regulation dynamics and to compare the performances for each system [10].

Also, some tasks related to frequency regulation in power systems are shown in articles [19, 20, 24, 25, 28–30].

Rapid development of renewable energy under conditions of large power systems, such as the power system of Ukraine, with an insufficient amount of maneuverable capacities and dominating base generation (with large blocks of thermal and nuclear power plants) requires a separate consideration. Significant intermittent capacities in such power systems with rapidly changing power output schedule can lead to frequency deviations and emergency situations. Taking into account the requirements for electricity quality in the field of frequency stability at a given level of 50 ± 0.2 Hz for the Ukrainian power systems and 50 ± 0.02 Hz for the EU power system ENTSO-E, this work focused on the development of a new model of frequency control processes, which would take into account these specific conditions and consider the possibility of using BESS as a regulator for similar power systems with large volumes of intermittent generation.

3 Mathematical Model

The mathematical model of this problem is a modified system of algebraic and differential equations and limitations, developed earlier in the study of frequency and power regulation processes in power systems in emergency modes with the help of consumers-regulators [1, 2]. The difference is in replacement of the equations describing the processes in the consumers-regulators (controllable loads) with the necessary equations for BESS.

The mathematical model obtained in this way has the form (1)—(16).

$$\frac{d\omega(t)}{dt} = \frac{\sum_{i=1}^{I} P_{pi}(t) + P_{BESS}(t) + P_{WPP}(t) + P_{SPP}(t) - P_l(t) - P_g(t)}{T_s P_{\Sigma p0}\omega(t)} - \omega_0^2, \quad (1)$$

$$\frac{dP_{pi}(t)}{dt} = \frac{P_{p0i} - P_{pi}(t) + B_{pi}(\omega(t) - \omega_0)}{\tau_{pi}}, \quad i = \overline{1, I}, \quad (2)$$

$$\frac{dP_l(t)}{dt} = \frac{P_{l0} - P_l(t) + C_l(\omega(t) - \omega_0)}{\tau_l}, \quad (3)$$

$$\frac{dP_{WPP}(t)}{dt} = \frac{P_{WPP0} + B_{WPP}(\omega(t) - \omega_0) + P_{WPPw}(v_w(t)) - P_{WPP}(t)}{T_{WPP}}, \quad (4)$$

$$P_{WPPw}(v_w) = c_0 + c_1 v_w + c_2 v_w^2 + \ldots + c_n v_w^n, \quad (5)$$

$$P_{SPP}(t) = \frac{1}{2}A_0 + \sum_{k=1}^{M}(A_k \cos k\omega_0 t + B_k \sin k\omega_0 t), \quad (6)$$

$$P_{BESS}(t) = P_{BESS0} + A_{BESS}(\omega_0 - \omega(t)) - Q_{BESS}\frac{d\omega(t)}{dt}$$

$$+ S_{BESS} \int_{t_0}^{t_1} (\omega_0 - \omega(\tau))d\tau, \tag{7}$$

$$v_w(t) = \frac{1}{2} A_{v0} + \sum_{k=1}^{N} (A_{vk} \cos k\omega_0 t + B_{vk} \sin k\omega_0 t), \tag{8}$$

$$A_{vk} = \frac{1}{N} \sum_{n=0}^{2N} \left(v_w(t_n) \cos \frac{2\pi k}{T} t_n \right), \quad k = 0, 1, \ldots, N, \tag{9}$$

$$B_{vk} = \frac{1}{N} \sum_{n=1}^{2N} \left(v_w(t_n) \sin \frac{2\pi k}{T} t_n \right), \quad k = 0, 1, \ldots, N, \tag{10}$$

$$A_k = \frac{1}{M} \sum_{m=0}^{2M} \left(P_{SPP}(t_m) \cos \frac{2\pi k}{T} t_m \right), \quad k = 0, 1, \ldots, M, \tag{11}$$

$$B_k = \frac{1}{M} \sum_{m=1}^{2M} \left(P_{SPP}(t_m) \sin \frac{2\pi k}{T} t_m \right), \quad k = 0, 1, \ldots, M, \tag{12}$$

where T_c is time constant of the grid; $P_{\Sigma p0}$ is the total power of the generators at the initial time; $P_{pi}(t), P_l(t), P_{BESS}(t), P_{WPP}(t), P_{SPP}(t)$ are required variables-functions corresponding to power generators, load, BESS, WPP and SPP respectively; $P_g(t)$ is grid losses; $\tau_{pi}, \tau_l, T_{WPP}$ are generator, load and WPP time constants respectively; B_{pi}, C_l, B_{WPP} are steepness of frequency characteristics of generators, loads and WPP respectively; $A_{BESS}, Q_{BESS}, S_{BESS}$ are coefficients of gain of the proportional, differential and integral components of the law of regulation for BESS; t is time variable; $v_w(t_n)$ and $P_{SPP}(t_m)$ is tabular wind speed and SPP power values at points t_n and t_m respectively; N and M is the number of natural measurements of wind speed and SPP power over the entire time interval respectively; A_{vk}, B_{vk}, A_k, B_k are coefficients of discrete Fourier transform.

In addition to Eqs. (1–12), the mathematical model includes the following constraints:

• limitations on power change rate

$$L_{BESSlg} \leq \left| \frac{d P_{BESS}(t)}{dt} \right| \leq L_{BESSug}, \quad t \in [t_0, T], \quad P_{BESS} \in [P_{BESSg1}, P_{BESSg2}], \tag{13}$$

- BESS power level limits

$$P_{BESS}.\min \leq P_{BESS}(t) \leq P_{BESS}.\max, \tag{14}$$

- dead band

$$\frac{dP_{BESS}(t)}{dt} = const, \ \omega(t) - \omega_0 \in [\omega_{s1}, \omega_{s2}], \tag{15}$$

where $[t_0,T]$ is time interval in which the processes in the power system are investigated; g is index indicating the BESS power interval in which the speed limit applies; $P_{BESS.min}$, $P_{BESS.max}$ are limits of minimum and maximum power of the BESS; $[\omega_{s1}, \omega_{s2}]$ is dead band of BESS.

Initial conditions:

$$\begin{cases} \omega(t_0) = \omega_0, \\ P_{pi}(t_0) = P_{pi0}, \\ P_l(t_0) = P_{l0}, \\ P_{WPP}(t_0) = P_{WPP0}. \end{cases} \tag{16}$$

As a control function for the BESS in this model was used the proportional-differential-integral (PID) law of frequency-dependent regulation (7).

4 Modelling Example

Model (1–16) have been used to calculate the process of frequency and power stabilization in the power system, in which, together with traditional power plants, powerful WPP, SPP and BESS operate in conditions of frequent change of power load of the power system. The aggregated calculation scheme of the studied power grid is shown in Fig. 5. The power system includes traditional power plants such as nuclear power plant (NPP), thermal power plant (TPP) and hydro power plant (HPP).

Values of necessary parameters for model calculations (1–16) according to the scheme in Fig. 5, are given in Table 1. The maximum power of the BESS is 3000 MW, the power change rate is 714% of the installed power per second.

The graph of wind speed, which was used during the study in the range of 0–600 s, was constructed using the Fourier transform on the basis of initial natural data [31] pattern with length of 135 s and is shown in Fig. 6. The obtained graph of the wind power plant output, which was calculated in the model (1–16) based on the generated wind speed are shown in Fig. 7.

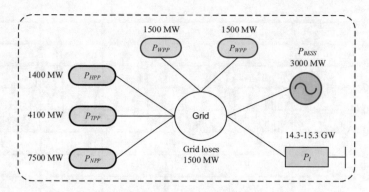

Fig. 5 Aggregated power system calculation scheme

Table 1 Parameters of the power system model with WPP, SPP and BESS according to Fig. 5

System element	Steepness of frequency characteristic, MW/Hz	Time constant, s	Initial power, MW
HPP	0.37	5	1400
TPP	0.73	5	4100
NPP	1.81	5	7500
Load	153	0.5	14300
BESS	–	–	342
WPP	0.4	5	1035
SPP	0.4	5	1423
Losses	–	–	1600

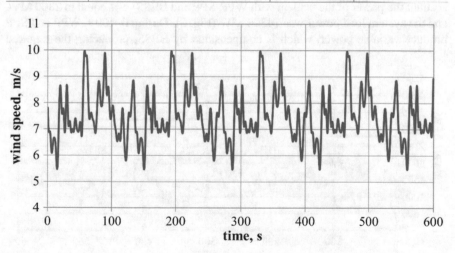

Fig. 6 Wind speed graph

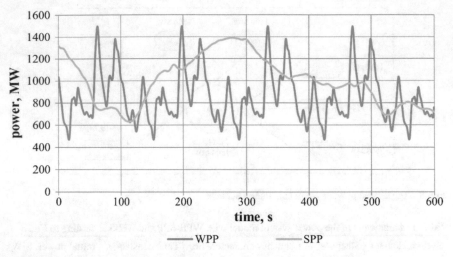

Fig. 7 Power output graph of WPP and SPP

The SPP power output is the analytical dependence obtained using the Fourier transform of the source data given in [4]. The graph of SPP output power is shown in Fig. 8.

The study simulated the operation mode of the power system with variable load capacity. Along with this, it was considered that WPP, SPP and BESS worked together as one system for power output.

At the initial point of time t = 0 s, the power imbalance (if we do not take into account the power of the tandem with WPP, SPP and BESS) was equal to 2800 MW, and the system load power was 14300 MW (Fig. 8). During 0–100 s, WPP and SPP produce variable power, which is compensated by BESS, producing the required

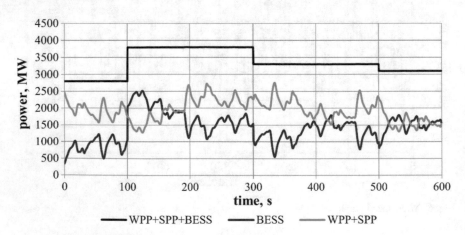

Fig. 8 Power of BESS, WPP + SPP and their total power

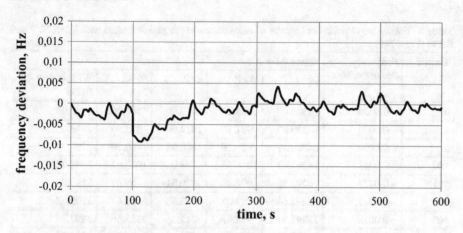

Fig. 9 Frequency deviation in the power system when WPP, SPP and BESS work together

amount of energy according to the value of imbalance, which allows to maintain the frequency with high accuracy (Fig. 9).

At the moment t = 100 s, an additional load of 1000 MW is added to the power system, which changes the total load power up to 15300 MW, at the same time the required tandem power is changed and taking the value of 3800 MW. Due to its high performance, the BESS quickly increases its power to compensate for increased load and keeps the frequency deviations at a level not exceeding 0.01 Hz. With this load, the system continues to work for 200 s, accurately maintaining the frequency at a given level.

After the next 200 s, the load power is changed again, but now in the other direction: the load power is reduced by 500 MW to 14800 MW and the required power of the WPP, SPP and BESS group of power plants is reduced to 3300 MW for imbalance coverage respectively.

BESS changes its power according to the received difference between the power of WPP, SPP and the required power of the tandem, keeping the frequency deviation within the range of 0.005 Hz and maintaining it at this level until the next change of the dispatcher's task.

At the moment t = 500 s the load of the power system decreases additionally by 200 MW with the corresponding change of the total load to 14600 MW and the required power of tandem 3100 MW.

The frequency deviation caused by the switching operation does not exceed 0.003 Hz. The load remains constant up to the time stamp t = 600 s, the frequency stabilization by BESS remains at the same high level.

The accuracy of frequency stabilization indicates that the total power of WPP, SPP and BESS corresponds to the lack of power in the system (see Table 2), which is achieved due to the high speed of the BESS and the effectiveness of the selected regulation law.

Table 2 Fragments of numerical results of calculations in the power system with WPP, SPP and BESS (Figs. 7–9)

Time, s	Frequency deviation, Hz	Power, MW				
		Load	WPP	SPP	BESS	WPP + SPP + BESS
0	0.0000	14300.00	1035.00	1423.00	342.00	2800.00
1	−0.0003	14299.87	991.17	1414.90	392.56	2798.63
2	−0.0006	14299.57	929.29	1408.93	460.10	2798.32
100	−0.0023	14297.87	1016.55	738.06	1042.62	2797.22
101	−0.0077	15293.33	996.96	730.31	2065.97	3793.23
102	−0.0077	15292.66	987.63	723.52	2081.47	3792.62
103	−0.0078	15292.54	971.42	717.36	2103.46	3792.24
104	−0.0080	15292.42	940.62	711.49	2139.76	3791.87
300	−0.0007	15299.37	782.00	1482.34	1535.15	3799.49
301	0.0021	14801.58	823.07	1477.45	1002.23	3302.76
302	0.0025	14802.17	899.62	1470.93	932.79	3303.35
303	0.0026	14802.46	939.38	1463.30	899.70	3302.38
304	0.0025	14802.45	920.04	1455.14	926.27	3301.46
500	0.0018	14801.86	1350.91	961.58	988.23	3300.72
501	0.0026	14602.48	1307.40	952.93	841.17	3101.49
502	0.0024	14602.38	1284.57	943.40	873.87	3101.85
503	0.0023	14602.26	1274.41	932.69	894.56	3101.67
504	0.0021	14602.09	1246.63	920.71	933.59	3100.93
597	−0.0012	14598.91	674.83	784.53	1639.17	3098.53
598	−0.0012	14598.83	666.08	780.41	1652.45	3098.94
599	−0.0010	14598.89	706.72	776.38	1617.29	3100.39

5 Conclusions

The results obtained confirm the possibility of using the developed mathematical model (1–16) to study the processes of frequency and power regulation in integrated power systems where simultaneously operate both wind and solar power plants together with battery energy storage systems.

The use of large BESSs in power systems as regulating capacities will allow to implement significant amounts of renewable energy sources based on WPP and SPP. This is can be achieved due to the high speed of BESS, which in combination with sufficient power (not exceeding the installed power of intermittent generation) allows to stabilize the frequency in the power system at acceptable level for normal operation and avoid the problem of unstable generation of renewable energy sources.

The high accuracy of frequency maintenance, which exceeds the regulatory requirements for the power systems of Ukraine and the European Union ENTSO-E, even when changing the set total power of units WPP, SPP and BESS, confirms the effectiveness of the used regulation law.

References

1. Kulyk, M.M., Dryomin, I.V.: General-puropse model of frequency and capacity regulation in united power systems. Prob. Gen. Energy **4**(35), 5–15 (2013)
2. Kulyk, M.M., Dryomin, I.V.: Generalized mathematical model and features of adaptive automatic frequency and power control systems. Prob. Gen. Energy **4**(43), 14–23 (2015). https://doi.org/10.15407/pge2015.04.014
3. Electricity and Energy Storage (Updated January 2020). World Nuclear Assotiation. Access mode: http://www.world-nuclear.org/information-library/current-and-future-generation/electricity-and-energy-storage.aspx
4. Marcos, J., Marroyo, L., Lorenzo, E., Alvira, D., Izco, E.: Power output fluctuations in large scale pv plants: one year observations with one second resolution and a derived analytic model. Prog. Photovol. Res. Appl. **19**(2), 218–227 (2010). https://doi.org/10.1002/pip.1016
5. Anzalchi, A., Pour, M.M., Sarwat, A.: A combinatorial approach for addressing intermittency and providing inertial response in a grid-connected photovoltaic system. In: Paper presented at the IEEE Power and Energy Society General Meeting, 2016-November (2016). https://doi.org/10.1109/pesgm.2016.7742056
6. Arani, M.F.M., El-Saadany, E.F.: Implementing virtual inertia in DFIG-based wind power generation. IEEE Trans. Pow. Sys. **28**(2), 1373–1384 (2013). https://doi.org/10.1109/TPWRS.2012.2207972
7. Argyrou, M.C., Spanias, C., Marouchos, C.C., Kalogirou, S.A., Christodoulides, P.: Energy management and modeling of a grid-connected BIPV system with battery energy storage. In: Paper presented at the 2019 54th International Universities Power Engineering Conference, UPEC 2019—Proceedings (2019). https://doi.org/10.1109/upec.2019.8893495
8. Arul, P.G., Ramachandaramurthy, V.K.: Mitigating techniques for the operational challenges of a standalone hybrid system integrating renewable energy sources. Sus. Energy Technol. Assess. **22**, 18–24 (2017). https://doi.org/10.1016/j.seta.2017.05.004
9. Asghar, F., Talha, M., Kim, S. H.: Fuzzy logic-based intelligent frequency and voltage stability control system for standalone microgrid. Int. Trans. Elec. Energy Sys. **28**(4) (2018). https://doi.org/10.1002/etep.2510
10. Bahloul, M., Khadem, S.K.: Design and control of energy storage system for enhanced frequency response grid service. In: Paper presented at the Proceedings of the IEEE International Conference on Industrial Technology, 2018-February, pp. 1189–1194 (2018) https://doi.org/10.1109/icit.2018.8352347
11. Bisht, M.S., Sathans: Fuzzy based intelligent frequency control strategy in standalone hybrid AC microgrid. In: Paper presented at the 2014 IEEE Conference on Control Applications, CCA 2014, pp. 873–878 (2014). https://doi.org/10.1109/cca.2014.6981446
12. Cho, S., Jang, B., Yoon, Y., Jeon, W., Kim, C.: Operation of battery energy, storage system for governor free and its effect. Trans. Korean Ins. Elec. Eng. **64**(1), 16–22 (2015). https://doi.org/10.5370/KIEE.2015.64.1.016
13. Habib, M., Ladjici, A.A., Harrag, A.: Microgrid management using hybrid inverter fuzzy-based control. Neu. Comput. Appl. (2019). https://doi.org/10.1007/s00521-019-04420-5
14. Guney, M.S., Tepe, Y.: Classification and assessment of energy storage systems. Renew. Sustain. Energy Rev. **75**, 1187–1197 (2017). https://doi.org/10.1016/j.rser.2016.11.102

15. Hao, X., Zhou, T., Wang, J., Yang, X.: A hybrid adaptive fuzzy control strategy for DFIG-based wind turbines with super-capacitor energy storage to realize short-term grid frequency support. In: Paper presented at the 2015 IEEE Energy Conversion Congress and Exposition, ECCE 2015, pp. 1914–1918 (2015). https://doi.org/10.1109/ecce.2015.7309930

16. Hassan, S.Z., Li, H., Kamal, T., Awais, M.: Stand-alone/grid-tied wind power system with battery/supercapacitor hybrid energy storage. In: Paper presented at the Proceedings of 2015 International Conference on Emerging Technologies, ICET 2015 (2016). https://doi.org/10.1109/icet.2015.7389179

17. Indu, P.S., Jayan, M.V.: Frequency regulation of an isolated hybrid power system with super-conducting magnetic energy storage. In: Paper presented at the Proceedings of 2015 IEEE International Conference on Power, Instrumentation, Control and Computing, PICC 2015 (2016). https://doi.org/10.1109/picc.2015.7455752

18. Jamroen, C., Dechanupapritta, S.: Coordinated control of battery energy storage system and plug-in electric vehicles for frequency regulation in smart grid. In: Paper presented at the 2019 IEEE PES GTD Grand International Conference and Exposition Asia, GTD Asia 2019, pp. 286–291 (2019). https://doi.org/10.1109/gtdasia.2019.8715962

19. Kulyk, M.M., Kyrylenko, O.V.: The state and prospects of hydroenergy of Ukraine. Techn. Electrodyn. **4**, 56–64 (2019). https://doi.org/10.15407/techned2019.04.056

20. Lytvynchuk, V.A., Kaplin, M.I., Bolotnyi, N.P.: The method of design an optimal under-frequency load shedding scheme. In: IEEE 6th International Conference on Energy Smart Systems, pp. 14–17 (2019). https://doi.org/10.1109/ess.2019.8764241

21. Melo, S.P., Brand, U., Vogt, T., Telle, J.S., Schuldt, F., Maydell, K.V.: Primary frequency control provided by hybrid battery storage and power-to-heat system. Appl. Energy **233–234**, 220–231 (2019). https://doi.org/10.1016/j.apenergy.2018.09.177

22. Naranjo Palacio, S., Kircher, K.J., Zhang, K.M.: On the feasibility of providing power system spinning reserves from thermal storage. Energy Build. **104**, 131–138 (2015). https://doi.org/10.1016/j.enbuild.2015.06.065

23. Orihara, D., Saitoh, H.: Evaluation of battery energy storage capacity required for battery-assisted load frequency control contributing frequency regulation in power system with wind power penetration. IEEJ Trans. Pow. Energy **138**(7), 571–581 (2018). https://doi.org/10.1541/ieejpes.138.571

24. Ponciroli, R., Passerini, S., Vilim, R.B.: Overview of simulation and control tools for extended operability of nuclear reactors. In: Paper presented at the 10th International Topical Meeting on Nuclear Plant Instrumentation, Control, and Human-Machine Interface Technologies, NPIC and HMIT 2017, pp. 1607–1616 (2017)

25. Shi, L., Xu, H., Li, D., Zhang, Z., Han, Y.: The photovoltaic charging station for electric vehicle to grid application in smart grids. In: Paper presented at the ICIAFS 2012—Proceedings: 2012 IEEE 6th International Conference on Information and Automation for Sustainability, pp. 279–284 (2012). https://doi.org/10.1109/iciafs.2012.6419917

26. Wen, Y., Dai, Y., Zhou, X., Qian, F.: Multiple roles coordinated control of battery storage units in a large-scale island microgrid application. IEEJ Trans. Elec. Elec. Eng. **12**(4), 527–535 (2017). https://doi.org/10.1002/tee.22408

27. Xie, P., Li, Y., Zhu, L., Shi, D., Duan, X.: Supplementary automatic generation control using controllable energy storage in electric vehicle battery swapping stations. IET Gen. Trans. Dis. **10**(4), 1107–1116 (2016). https://doi.org/10.1049/iet-gtd.2015.0167

28. Babak, V.P., Babak, S.V., Myslovych, M.V., Zaporozhets, A.O., Zvaritch, V.M.: Technical provision of diagnostic systems. In: Diagnostic Systems For Energy Equipments. Studies in Systems, Decision and Control, vol 281, pp. 91–133. Springer, Cham (2020). https://doi.org/10.1007/978-3-030-44443-3_4

29. Yan, S., Wang, M., Yang, T., Hui, S.Y.R.: Instantaneous frequency regulation of microgrids via power shedding of smart load and power limiting of renewable generation. In: Paper presented at the ECCE 2016—IEEE Energy Conversion Congress and Exposition, Proceedings (2016). https://doi.org/10.1109/ecce.2016.7855207
30. Yang, T., Mok, K., Ho, S., Tan, S., Lee, C., Hui, S.Y.R.: Use of integrated photovoltaic-electric spring system as a power balancer in power distribution networks. IEEE Trans. Pow. Elec. **34**(6), 5312–5324 (2019). https://doi.org/10.1109/TPEL.2018.2867573
31. Kolesnikov, A.: What is the wind? Access mode: http://al-kolesnikov.livejournal.com/17152. html

Energy Efficient Renewable Feedstock for Alternative Motor Fuels Production: Solutions for Ukraine

Anna Yakovlieva⬡ and Sergii Boichenko⬡

Abstract The presented study is devoted to assessment of renewable feedstock for alternative fuels production with relation to quality parameters, agrotechnical parameters, environmental impact and overall sustainability of possible feedstock. The state-of-art of the modern transport sector is shown, as well as the need in fostering implementation of biofuels for road and air transport. Rapeseed and camelina oils are considered as two of the most profitable sources for biofuels production in Ukraine. The comparative analysis of quality and agrotechnical parameters was done. Taking into account requirements to sustainability it is recommended to develop cultivation of camelina in Ukraine as a feedstock for biofuels production.

Keywords Motor fuel · Jet fuel · Biofuel · Renewable feedstock · Rapeseed oil · Camelina oil · Environment

1 Introduction

The rapid development of the world economy is accompanied by high rates of primary energy consumption. In recent years, global consumption of major energy sources of industrial value has exceeded 10 billion tons/year [1].

The significant increase in number of road transport in recent years has led to a great demand for petroleum products. However, oil reserves are estimated to last only a few decades. As a consequence, depletion of oil reserves will have a significant impact on the transport sector. In this regard, today in the world there is an active search for alternative fuels [1]. The use of biofuels instead of conventional fuels is a rather urgent problem for our country, in particular, since Ukraine belongs to energy-scarce countries and has relatively low oil and gas reserves.

The main consumer of fuel worldwide is road transport (2575 million as of 2017) [9]. In the structure of road transport of Ukraine, cars with gasoline engines (more

A. Yakovlieva (✉) · S. Boichenko
Department of Chemistry and Chemical Technology, National Aviation University, Kyiv, Ukraine
e-mail: anna.yakovlieva@nau.edu.ua

© The Editor(s) (if applicable) and The Author(s), under exclusive license   247
to Springer Nature Switzerland AG 2020
V. Babak et al. (eds.), *Systems, Decision and Control in Energy I*, Studies in Systems,
Decision and Control 298, https://doi.org/10.1007/978-3-030-48583-2_16

than 85%) are predominant, there are about 13% of diesel vehicles and less than 1.5% of gas tanks [9].

Transport as a sector of the economy is one of the most powerful factors of anthropogenic impact on the environment. Some types of this impact, such as air pollution and noise pollution, are among the most serious man-made loads on environmental components of particular regions, especially large cities [2, 9].

With increasing concern about reducing fossil energy resources and minimizing environmental footprint, there is increasing interest in exploring alternative energy sources, including alternative motor fuels or biofuels.

Today, one of the promising areas of research in the field of transport and its related fuel supply is the development of alternative environmentally friendly motor fuels from renewable raw materials of plant origin. The use of such fuels will help reduce the anthropogenic load on the environment, reduce the amount of CO_2 emissions into the atmosphere (both in the process of fuel production and their use). In addition, it will reduce the dependence of a number of countries on imports of petroleum products and other energy resources [12, 13].

Given the rapid development of technology, there is a wide variety of alternative motor fuels in the world today. In particular, biofuels have become widespread; primarily due to the ability to overcome the ongoing climate change caused by carbon dioxide emissions and reduce our dependence on oil resources. Over the last decades, scientists have been exploring the use of various types of biomass for biofuels.

2 Generations of Alternative Motor Fuels

The search and widening of new feedstock, development of progressive technologies of alternative fuels production and their rational use in transport, is one of the priority tasks nowadays. The advantages of renewable energy are: natural origin, rapidity of renewal, absence of extra CO_2 emissions, less negative impact on environment, easy biodegradation in nature. At the same time alternative fuels from renewable feedstock must meet the requirements connected to efficiency, reliability and durability of transport [5–8].

Today there is a great variety of renewable feedstock or biomass for alternative fuels production [4]. Biomass-derived alternative fuels can provide a near-term and even a long-term solution to the transport industry with a lower environmental impact than petroleum fuels.

Potential feedstock for producing alternative fuels are classified as:

- *oil-based feedstocks*, such as vegetable oils, waste oils, algal oils, and pyrolysis oils,
- *solid-based feedstocks*, such as lignocellulosic biomass (including wood products, forestry waste, and agricultural residue) and municipal waste (the organic portion),
- *gas-based feedstocks*, such as biogas and syngas.

The key to the successful implementation of alternative fuels is the availability of feedstock at a large and sustainable scale and low price. Improved yields and reduced plantation or transportation costs would promote commercialization of alternative fuels conversion processes and, therefore, would allow the industry and government to assess and address the feedstocks' potential and impacts [4, 13–15].

It is obvious that each kind of feedstock requires certain technology of its processing. Some of them are already popular and some are still developing. According to the complexity and the level of maturity these technologies are traditionally classified according to generations (Fig. 1).

First generation of alternative fuels (biofuels), which are widely popular today, are produced from traditional agricultural plants. Plant oil-derived biofuels are produced by esterification of traditional oil-containing plants (rape, soya, sunflower, palm, etc.). Ethanol is produced by fermentation of the sugar or starch contained in plant biomass.

Second generation of biofuels is produced by processing the whole plant—particularly its lignocelluloses, the main component of plant cell walls. The resource is available in large quantities in a variety of forms: wood, straw, hay, forestry waste, plants' residues, etc. second generation processes do not compete with food uses. Two processes are being studied: biochemical conversion and thermochemical conversion [3, 13].

Third generation of biofuels can be produced using either autotrophic (operating via photosynthesis) algal biomass or heterotrophic process (operating via the supply of an external carbon, such as sugar). Except that third generation biofuels foresee applying the already known technologies for new kinds of oil-containing feedstock, which do not compete with food industry (jatropha, camellina, etc.) [15, 16].

Fourth generation biofuels (Fig. 2) are considered only during the last 3–5 years and now still at the early stage of development. They are derived from specially engineered biomass, which has higher energy yields or are able to be grown on non-agricultural land or water bodies. Biomass crops are seen as efficient 'carbon capturing' machines that take CO_2 out of the atmosphere and 'lock' it in their branches, trunks and leaves. The carbon-rich biomass is converted into fuels by means of second generation techniques. The greenhouse gas is then geosequestered—stored in depleted oil and gas fields, in unmineable coal seams or in saline aquifers, where it stays locked up for hundreds of years [17, 18].

3 Perspective Plant-Derived Components for Alternative Jet Fuels Production

This diversity of technologies provides the ability for alternative fuels production using various feedstocks. Scientists believe that the most promising feedstocks are plants with high oils content, algae and some types of industrial and household waste.

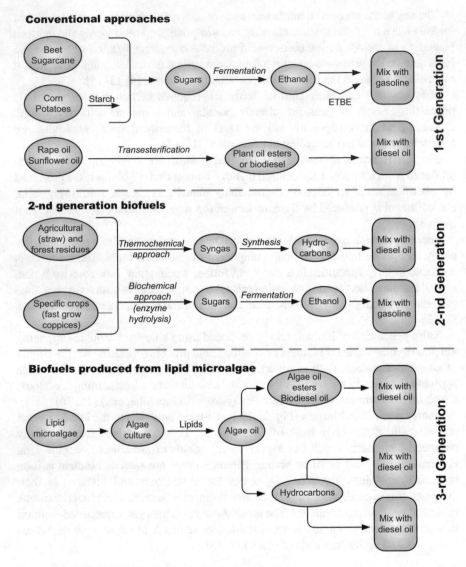

Fig. 1 Development of technologies for biofuels production

Camelina relates to energy crops with high oil content. The main consumers are the producers of biofuels. Camelina cultivation is currently implemented at sufficiently large scale so to achieve a considerable production of crude vegetable oil that will then be converted into paraffinic biofuels (HEFA). Camelina is used in agriculture as crop rotation, which prevents reduction of soil fertility and provides increasing of crop resistance to diseases and pests. It is not demanding to climatic conditions and does not require substantial cultivation and care. Camelina seeds contain 40–50% of

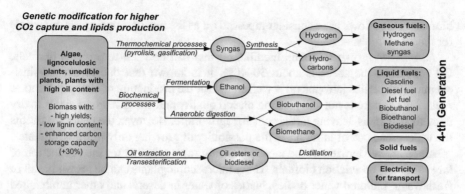

Fig. 2 Technologies for 4th generation alternative fuels production

oil, providing oil output of about 1250 l/ha. Another advantage of this culture is the use of meal as feed for livestock and poultry. The scientists believe these camelina characteristics provide "sustainability" of the process of biofuels production without creating competition in the food industry. Camelina is a very promising crop that is expected to offers a high level of sustainability, that can be grown in EU and elsewhere in marginal land where conventional agriculture is not sufficiently productive to be carried out by farmers.Nowadays this culture is widespread in the US, Canada and some European countries [3, 4].

Rape is considered to be one of the main crops for biofuels during the last 10–15 years. During 2000–2010 years leading producers of rapeseed oil were Canada, the US and European countries such as Germany, France, Czech Republic, Poland, UK. The chemical composition and basic characteristics of rapeseed oil are well suited for alternative fuels. But then the question arose about the necessity of rape cultivation as the main biofuels feedstock. Rape culture is highly depending on growing conditions, needs constant fertilizing and other care while significantly depletes the soil in areas that are traditionally used for agriculture. Analyzing these data, scientists have concluded that rape is competitive in needs of the food industry, and thus the process of production and usage of biofuels will not be sustainable [3]. Numerous researches on selection rapeseed crops with improved physicochemical and agronomic characteristics were held. Thus, the oil content in the seed yield was increased as well as average oil output to 1200–1300 l/ha. The so-called low-erucic breeds with the corresponding fatty acid composition of the rapeseed oil were selected that is the best for biofuels. The scientists were able to increase crop resistance to pests, climatic conditions or other adverse growing conditions, and thus reduce the cost of oil production and fuel respectively. We can conclude that the denial of rapeseed as a feedstock for biofuels is not fully justified [9, 10].

It should be noted that for Ukraine rape is a typical culture, and in the last 8–10 years the massive producing was noticed. The most rapid increase of production volumes of rape occurred in the period 2004–2008 years, which coincided with the growing of global demand for rapeseed oil as raw material for the production of biofuels. This allows us to consider rapeseed oil as the most promising raw materials for motor biofuels.

Jatropha is grown to obtain inedible oil that can be used to produce biofuels. Oil content in the seeds is about 30–40%. It is known that due to the physical-chemical properties jatropha oil is well suited for the production of alternative motor fuels. This culture is not a depending on soil quality and can give good yields in dry uncultivated areas, leaving fertile soils for agriculture. However, the factor that limits the widespread use of jatropha, is the possibility of growing only in warm climates.

Algae are recognized as the most promising raw materials for the production of large amounts of aviation biofuels. These microscopic plants can be grown in salt or fresh water, polluted water bodies, bodies of water in deserts and other uninhabited areas. During the lifecicle algae consume CO_2, which makes them extremely effective tool for the absorption of carbon from the atmosphere produced from burning of fossil fuels. Microalgae are capable for producing up to 15 times more oil on 1 km^2 than other energy crops. The can grow in low-boundary land not used for farming (desert areas).

It might be feasible to convert algae oil directly into green diesel or aviation fuel. Catalytic conversion of second generation oil with hydrogen into paraffins has already been investigated for application in industry.

The most common liquid fuel from algae is Fatty Acid Methyl Ester (FAME) is typically produced by a reaction (transesterification) between triglycerides and alcohol (most commonly methanol) at 60–70 °C and in the presence of an alkaline or acidic homogeneous catalyst at atmospheric pressure. In addition to triglycerides in the lipid fraction, algae oil also contains a substantial quantity of free fatty acids and some moisture. Their occurrence is undesirable for transesterification because in alkaline catalysed reaction, they produce soap and reduce biodiesel yield. In this instance, an acid catalyst is better suited for the purpose as it is able to process low grade feed. Current processes used for manufacturing biodiesel are not entirely suitable for algae oil. A feasible option is to carry out the reaction at high temperature and pressure. The supercritical transesterification of first and second generation oil by various research groups showed almost complete conversion in reasonable reaction time. At supercritical conditions, the reaction can process moisturerich feed with free fatty acids and subsequently eliminate pre-treatment process units. Application of catalyst-free supercritical alcohol transesterification is desirable due to feed stock flexibility and the relatively small reactor volume needed to achieve high production rates.

One of the main advantages of using algae is their massive biodiversity, which makes it possible to select strains for a particular geographical location or a specific purpose. Different algal strains have adapted to grow in UK soil, on the surface of the ocean, underneath desert sand, next to hydrothermal vents, and in freezing Siberian rivers. There is an optimal algal strain for every location. One potential algal application is to capture the CO_2 emitted by fossil fuel combustion in power stations. To achieve this, it is necessary to select an extremophile with high temperature and low pH tolerance, as well as a very active Calvin Cycle.

However, at this stage of development the issue of the best technologies development for algae cultivation is not yet fully resolved. Scientists project this process requires at least 8–10 years.

Halophytes are herbaceous plants that are combined in a separate group due to the possibility of their growth on saline soils (salt marshes and other areas with access to sea water). The using of this type of material is very promising in terms of the idea of sustainability of alternative motor fuels. Today about 20–25% of the Earth is not used for agriculture through increased soil salinity levels. Typical examples of halophytes are rich in unsaturated fatty acids containing 90% of carbon chains in triglycerides. Overall cultivation of halophytes is an important component of the system designed to reduce the amount of greenhouse gases. The use of these plants as a feedstock for the production of aviation biofuels is still at the research stage, but could be widely implemented in 2–4 years.

Household and industrial wastes in recent years are also considered a promising raw material for production of aviation biofuels. Currently factories are actively building, where as a result of complex processes such as waste wood, paper, wood residues, agricultural residues, by-products of livestock, some industrial waste, food waste, municipal sewage and others are processing into fuels. One of the advantages of the use of waste is the ability to ensure the production of biofuels from waste plant material. In addition, recycling of waste into alternative fuels is one of the solutions to the problem of waste recycling that accumulates on numerous landfills.

Used cooking oil (UCO) is another feedstock that is being investigated. Compared to other more traditional oily feedstock, UCO shows a high variability in quality and composition depending on the collection area, the collection method, and the period of the year (when different vegetable oils are consumed). Contaminations in the oil represent a serious technical challenge to be dealt with before catalytic hydrotreatment process [10–12].

The upgrade of crude UCO to a suitable feedstock is an activity that still requires R&D work, and new approaches can be considered. This variability could be reduced by proper pre-treatments. Once collected, UCO is usually filtered and de-watered: water in UCO can create problems in the downstream processes, and is source of damages to plant equipment, enhancing the corrosion.

The water content of UCO is normally reduced by means of paper filters, a cost-effective solution with high removal efficiency. The use a paper filtration contextually reduces the amount of suspended solids. The total contamination has to be significantly reduced, since solids in aviation engines are cause of injector and blade corrosion and erosion, increasing the maintenance engines costs and risks.

Due to vegetable oil cooking processes, and contamination from the food, UCOs usually show high free acidity (1–5%). A common possible mean to improve the UCO quality, reducing acidity without losses in its energy content, is the esterification process. This process, widely used in the biodiesel sector, rebuilds the triglycerides by using glycerine in a dedicated reactor, sometimes in presence of catalysts. However, the interest in this technique depends on the type of downstream processes: this process is normally adopted in the biodiesel sector, where in most of the cases transesterification to biodiesel does not tolerate high level of fatty acids, and could also be of interest for HEFA production.

Today most of scientists claim about such property of alternative motor fuels as its sustainability. Sustainable motor fuels can be considered as sustainable only if they have a substantially better GHG balance than their fossil alternative, do not harm the environment, or involve any negative socio-economic impacts. Not all biomass feedstock are fit to produce sustainable motor fuels. The type and origin of the biomass feedstock largely determines the overall sustainability of the sustainable motor fuels, including the lifecycle of its GHG mainly through production and transport energy needs, use of fertilizers and land-use change effects. Some types of biomass feedstock may actually cause more GHG emissions than conventional fossil motor fuel especially when considering indirect land use change impacts [19–21].

Table 1 presents a summary of the emissions reduction for a number of feedstock. Sustainable motor fuels derived from wastes (such as animal fat and used cooking oil), or based on wood and agricultural residues (such as straw), have significantly lower emissions than those based on conventional oil crops. Sustainable motor fuels from algae can potentially be carbon neutral or even produce a reduction on GHG thanks to the absorption of CO_2 in co-products other than fuel. While these numbers are a useful guide, actual GHG reductions are ultimately dependent on the design of each specific project.

In many cases the cost competitiveness of biofuels therefore depends on the price of feedstock. The cost of feedstock includes the price of raw material and its eventual pre-treatments. Transport costs from the feedstock supplier to the alternative motor fuels plant must be added too. Table 2 shows a selection of the wide range of variables influencing feedstock, logistics and pre-processing costs.

Table 1 Greenhouse gas emissions of Sustainable motor fuels

Technology pathway	Feedstock	Emissions (gCO$_2$/MJ fuel)	Savings CO$_2$ versus conventional fuels (baseline)
	Conventional fuel (average value)	87.5	
FT	Wood residues/straw	4.8	95%
HEFA	Conventional oil crops (palm oil, soy, rapeseed)	40–70	20–54%
	Jatropha	30	66%
	Camelina	13.5	85%
	Animal fat	10	89%
	Algae (from open ponds)	− 21 (best case) 1.5 (realistic case)	124% (best case) 98% (realistic case)

Table 2 Variables influencing feedstock, logistics and pre-processing costs

Feedstock costs	Logistics and pre-processing costs
Geographic origin	Distance
Feedstock type	Accessibility
Seasonality (droughts) and availability	Mode of transportation
	Technology level
Level of mechanization and inputs	Scale
Scale	

Projects and strategies to increase feedstock yields and to optimize logistics in terms of availability and infrastructure are required to ensure the provision of feedstock at competitive prices.

4 Analysis of Renewable Feedstock Available in Ukraine

High-energy-intensive oil crops are the most attractive source of biodegradable feedstock, which can partially replace traditional crude oil, including in the production of aviation jet fuels. However, the use of traditional oils for technical purposes, in particular for the production of fuel, is significantly restricted. This is due to the fact that such crops as sunflower, soybean, corn, etc. are grown primarily to meet the needs of the food industry. For the same reasons, cultivation of industrial crops, such as rapeseed, is limited to the area available for cultivation. Today, one of the most promising alternative oilseeds, which is characterized by low requirements for growing conditions, is red sowing (Camelina sativa L.) [1, 2]. Camelina culture is very unpretentious to soil quality, climatic conditions, resistant to pests, diseases and cold, does not require a large amount of mineral fertilizers. In shorter duration of the growing season, this crop yields a proportion of rapeseed with the yield of oil per unit area of crop. Short growing season also contributes to the cultivation of camelina as an intermediate crop in post-harvest crops. In addition, growing camelina does not lead to intensive depletion of fertile land [4].

Camelina belongs to the cabbage family Camelina and includes 15 species, of which the most widely cultivated redhead sowing. It is considered to be the least demanding of growing conditions compared to other oilseeds.

Camelina is characterized by high cold resistance (the seed germinates beyond 1 °C, and the seedlings easily withstand freezing to minus 12 °C) and water-time drought resistance. It grows well on all types of soil except clay. One of the main biological features of Camelina is the short growing season, which in most regions of cultivation is 80–85 days (due to which it reaches, and can be successfully grown in all regions of Ukraine), which makes it possible not only to effectively use the moisture reserves in the fall—winter rainfall, but also to form a crop due to rainfall that occurs during the growing season [3, 4].

The short growing season of the Camelina enables it to grow other crops after harvesting, and using it for a busy couple allows it to prepare the soil well and accumulate moisture for winter sowing.

In addition, unlike other cultures of the cabbage family, it is practically not pest-infested and disease-free, and in times of steady increase in energy and pesticide prices, it can significantly reduce the cost of growing it. Camelina is quite a crop crop: its potential yield can be 20–30 centners/ha. Camelina seeds contain more than 40–50% of oil and 25–32% of crude protein. Unsaturated fatty acids with several double bonds predominate in redhead oil. Due to this, the oil has a rather low pour point—about mines 18 °C, which will subsequently provide satisfactory low temperature properties of biofuels.

One of the main advantages of Camelina oil is that the process of growing it does not require much cost, which is reflected in the cost of the oil.

In order to substantiate the expediency of growing Camelina oil as an alternative raw material for the production of biodiesel, a comparative analysis of the basic agrotechnical characteristics of the Camelina and rapeseed crops was conducted. The results of the analysis are shown in Table 3.

After analyzing data in Table 3, it can be concluded that Camelina has more potential than rapeseed, primarily due to its resistance to adverse soil and climatic conditions. As a result, it can be grown on low quality soils that are not suitable for other crops.

In addition, the cultivation of Camelina is environmentally safe, because Camelina is characterized by extreme plasticity to agro-ecological conditions of cultivation, does not require the use of fertilizers, pesticides and fungicides [13].

Table 3 Comparative characteristics of Rapeseed and Camelina cultures

Properties	Camelina	Rapeseed
General characteristic	One-year culture	One-year culture
Dry-resistance	Low water requirement	High water demand
Germination potential	Almost all types of soil are suitable	Demanding to the soil
The threat of declining soil fertility	Used as an intermediate crop, after which other crops can be planted	Depletes soils. It is possible to grow rapeseed in the same place only in 3–4 years
The presence of weeds	Isolated essential oil that inhibits the growth and development of weeds from the stage of stem formation to the full maturity of the seeds	A lot of weeds
Seed loss	High resistance to cracking	Low resistance to cracking
Vulnerability of crops by pests	Pests and diseases are not Detected	Strongly affected by pests
Vegetation period	60–75 days	90–100 days

From [1, 2], it is known that in Ukraine, nowadays, the sown area of Camelina occupies 5–6 thousand hectares (3% of all oilseeds), mainly in the northern part of the left bank of the Forest-Steppe. It is also noted that these areas can be increased by 3–4 times. However, to date, there is no clear strategy in Ukraine to develop Camelina cultivation [4].

Instead, in the US, Canada, some EU countries are considered to be one of the most promising oilseeds for biofuels production. It has been shown that until recently rapeseed was the main raw material for the production of alternative motor fuels (the so-called "first generation" biofuels). However, due to a number of difficulties and ambiguous issues related to rapeseed cultivation, the situation has changed significantly over the last 5 years. The reason for this is the considerable resource and energy costs of the rapeseed cultivation process. This is due to the fact that obtaining high rape crops requires intensive cultivation of fertile land, which quickly depletes them, the introduction of large quantities of fertilizers and pest control substances. Rape is very demanding on climatic and soil conditions. In addition, the problem is that rapeseed production on fertile land is a competition for food products. In this regard, further research into the use of rapeseed to produce biofuels, including aviation, is not appropriate.

An alternative solution to the problem is to use renewable plant material that does not compete with the food industry and does not present a short or long-term threat to the environment, i.e. the receipt and use of which is sustainable. This view is expressed by scientists in, calling alternative fuels on the basis of renewable raw materials second-generation biofuels. One of these types of raw materials, in particular, is Camelina oil.

A successful experience of using Camelina oil to produce aviation biofuels is described. Thus, flight tests were successfully carried out on airplanes of various types using mixtures of traditional fuel for jet engines of petroleum origin and products of processing of red oil. The latter were a mixture of aromatic hydrocarbons and C_7–C_9 isoalkanes, which were obtained by hydrodeoxygenation of oil, followed by selective cracking of the formed linear alkanes and isomerization and aromatization of the products of cracking. However, the main disadvantage of such biofuels is that, in chemical composition, they are a mixture of carbon-hydrogen, similar to petroleum jet fuels.

Another solution to the above problem is the potential for the introduction of fatty acid alkyl esters into the propellant composition. From works it is known that their production is based on the simplest chemical and technological point of view of the process of transformation of triglycerides—alcoholysis (esterification) with monohydric alcohols.

5 Conclusions

In view of the above, it is quite obvious that growing Camelina as a raw material for biofuel production will have some advantages over other crops used today. The use of biofuels based on Camelina oil will reduce CO_2 emissions throughout the lifecycle of such fuels, reduce the anthropogenic load on the environment and expand the raw material base for the production of motor fuels. This will reduce energy dependence on non-renewable energy sources.

References

1. Dangol, N., Shrestha, D.S., Duffield, J.A.: Life-cycle energy, GHG and cost comparison of camelina-based biodiesel and biojet fuel. Biofuels (2017). http://doi.org/10.1080/17597269. 2017.1369632
2. Chaturvedi, S., Bhattacharya, A., Khare, S.K., Kaushik, G.: *Camelina sativa*: An Emerging Biofuel Crop. Handbook of Environmental Materials Management, pp. 1–38. Springer (2018)
3. Solis, J.L., Berkemar, A.L., Alejo, Kiros, L.: Biodiesel from rapeseed oil (*Brassica napus*) by supported Li_2O and MgO. Int. J. Energy Env. Eng. **8**(1), 9–23 (2017). https://doi.org/10.1007/s40095-016-0226-0
4. Panchuk, M., Kryshtopa, S., Shlapak, L., Kryshtopa, L., Yarovyi, V., Sladkowski, A.: Main trend of biofuels production in Ukraine. Trans. Probl. **12**(4), 95–103 (2017). https://doi.org/10.20858/tp.2017.12.4.2
5. Issariyakul, T., Dalai, A.K.: Biodiesel production from greenseed canola oil. Energy Fuels **24**(9), 4652–4658 (2010). https://doi.org/10.1021/ef901202b
6. Dunn, R.O.: Alternative jet fuels from vegetable oils. Trans. ASAE **44**(6), 1751–1757 (2001)
7. Zaleckas, E., Makarevičienė, V., Sendžikien, E.: Possibilities of using Camelina sativa oil for producing biodiesel fuel. Transport **27**(1), 60–66 (2012)
8. Demirbas, A., Karslioglu, S.: Biodiesel production facilities from vegetable oils and animal fats. Energy Sourc. Part A: Recov. Util. Env. Eff. **29**(2), 133–141 (2007). https://doi.org/10.1080/009083190951320
9. JAL Biofuel Demo Flight First to Use Energy Crop Camelina [Електронний ресурс]/ CSR/ Environment, Tokyo.—December 16, 2008.—Режим доступу: http://press.jal.co.jp/en/release/200812/003149.html
10. Moser, B.R.: Camelina (Camelina sativa L.) oil as a biofuels feedstock: Golden opportunity of false hope? Lipid Technol. **22**(12), 270–273 (2010)
11. Wang, W.-Ch., Tao, L., Markham, J., Zhang, Y., at. al.: Review of Biojet Fuel Conversion Technologies **98** (2016)
12. Chuck, C.J., Donnelly, J.: The compatibility of potential bioderived fuels with Jet A-1 aviation kerosene. Appl. Energy **118**, 83–91 (2014)
13. Pearlson, M., Wollersheim, C., Hileman, C.J.: A techno-economic review of hydroprocessed renewable esters and fatty acids for jet fuel production. J. Biofuel. Bioprod. Biorefining **7**, 89–96 (2013)
14. Yakovlieva, A.V., Boichenko, S.V., Leida, K., Vovk, O.A., Kuzhevskii, Kh: Influence of rapeseed oil ester additives on fuel quality index for air jet engines. Chem. Technol. Fuels Oils **53**(3), 308–317 (2017)
15. Lapuerta, M., Canoira, L.: The suitability of Fatty Acid Methyl Esters (FAME) as blending agents in Jet A-1. Biofuels for Aviation. Feedstoc. Technol. Imp. **4**, 47–84 (2016)
16. Iakovlieva, A., Vovk, O., Boichenko, S., Lejda, K., Kuszewski, H.: Physical-chemical properties of jet fuels blends with components derived from rapeseed oil. Chem. Chem. Technol. **10**(4), 485–492 (2016)

17. Yakovlieva, A.V., Boichenko, S.V., Lejda, K., Vovk, O.O.: Modification of jet fuels composition with renewable bio-additives, p. 207. Kyiv, NAU (2019)
18. Cherian, G.: Camelina sativa in poultry diets: opportunities and challenges. Biofuel co-products as livestock feed—Opportunities and challenges, at Ed. Harinder P.S.M.—Rome, Ch. 17, pp. 303–310 (2012)
19. Iakovlieva, A.V., Boichenko, S.V., Vovk, O. O.: Overview of innovative technologies for aviation fuels production. Chem. Chem. Techn. **7**(3), 305–312 (2013)
20. Shkilniuk, I., Boychenko, S., Turchak, V.: The problems of biopollution with jet fuels and the way of achiving solution. Transport/S. Boychenko **23**(3), 253–257 (2008)
21. Aviation chemmotology: fuel for aviation engines. Theoretical and engineering bases of application: textbook/ N. S. Кulік, A. Ф. Aksionov, L. S. Yanovskiy, S. V. Boichenko, p. 560. Кyiv: NAU (2015)

Source Term Modelling for Event with Liquid Radioactive Materials Spill

**Yurii Kyrylenko⊙, Iryna Kameneva⊙, Oleksandr Popov⊙,
Andrii Iatsyshyn⊙, Volodymyr Artemchuk⊙, and Valeriia Kovach⊙**

Abstract Today more and more often radiological assessors use decision support systems (DSSs) and other calculation tools to provide modelling of radiological impact (JRODOS ARGOS, HPAC, TurboFRMAC, et al.). In addition to the primary purpuses, this software is applied in emergency planning in conjunction with quantitative and qualitative analysis of possible scenarios for the development of accidents. It is primarily explained by the extensive capabilities of the assessment tools, as in terms of design power as well as in terms of convenience of input data description, acquisition and analysis of results. However, there is currently a significant challenge for the users of these DSSs in preparing a correct and complete package of input data, in particular, in the part of the source term data for each of the emer-gency scenarios that are being considered. A large number of works is devoted to the development of individual special a means of creating an information link between the source term and radiation impact assessment tools. The need to consider a large variety of potential emergency processes at nuclear facilities energy at this stage significantly complicates the creation of a universal and holistic the source term preparation tool. This paper is focused on source term modelling for the group of events associated to spill of liquid radioactive material in area with forced ventialation. This work lights such issues as nature of source term, requirements from the side of atmospheric dispersion modelling taken into account, mathematical modelling of release formation, suggested ways of the model integration in radiological consequences toolkit

Y. Kyrylenko (✉)
State Enterprise "State Scientific and Technical Center for Nuclear and Radiation Safety", Kyiv, Ukraine
e-mail: uokyrylenko11235@gmail.com

Y. Kyrylenko · I. Kameneva · O. Popov · A. Iatsyshyn · V. Artemchuk
Pukhov Institute for Modelling in Energy Engineering of NAS of Ukraine, Kyiv, Ukraine

O. Popov · A. Iatsyshyn · V. Artemchuk · V. Kovach
State Institution "Institute of Environmental Geochemistry" of NAS of Ukraine, Kyiv, Ukraine

V. Kovach
National Aviation University, Kyiv, Ukraine

V. Babak et al. (eds.), *Systems, Decision and Control in Energy I*, Studies in Systems, Decision and Control 298, https://doi.org/10.1007/978-3-030-48583-2_17

261

and testing results. Atmospheric way of radioactive substances spreading is considered as one of important according to actual requirements from the side of safety analysis.

Keywords Source term modelling · Atmospheric release · Radioactive discharge · Decision support system · Tanks · Liquid radioactive material · Spill · Leakage · Radiological consequences · Probabilistic safety analysis

1 Introduction

In order to improve the safety of nuclear power plants, it is necessary to sophisticate the methodological and instrumental base for the analysis and safety assessment of currently operating nuclear power units, as well as the planned power units. The development of new computers and mathematical models for the estimation of radiation consequences is aimed at solving a lot of problems in this area [1, 2]. First of all, these are tasks such as:

- minimizing of radiation impact on the public, personal and the environment, taking into account social and economic factors (ALARA principle), in accordance with the Law of Ukraine "On Nuclear Energy Use and Radiation Safety", the national Radiation Safety Standards NRBU-97 [3] and IAEA safety requirements;
- analysis of normal operation, design basis accidents and late phases of beyond-design basis accidents at NPPs within the framework of the design documentations of the operating organization and environmental impact assessments [4, 5];
- expert assessment of safety analysis reports in accordance with current nuclear and radiation safety regulations, rules and standards;
- development of probabilistic safety analysis level 3 [5];
- emergency preparedness and response.

2 Literature Analysis and Problem Statement

In accordance with the Energy Strategy of Ukraine for the period up to 2035 [6] and the CCSUP Program [7], the state plans to increase the level of safety of the sites in the nuclear industry. There are many works available today that allow us to estimate the quantitative characteristics of an emergency releases from such sites and its radiological consequences [8, 9]. It is also worth noting the research of such scientists as M.N.A. Pereira [10], R. Schirru [10], S.S. Raja Shekhar [11], I.V. Kovalets [12], I.A. Ievdin [12], L. Robertson [12], A.R. Zabirov [13], A.A. Smirnova [13], L. Hu [14] et al. [1, 8, 15–19], whose works deals with the development of methods, algorithms, software and hardware for solving environmental and radiation safety problems. Since the late 1990s, major conservative approaches to the determination of source term parameters and radiation consequences have already found their place in the operational design of the operating organization and in the expertise of the regulatory body for a wide range of accident conditions.

However, in light of the probabilistic safety analysis for Ukrainian NPPs and NRBU-97/D-2000 [20] requirements for potential exposure of the population, in the late 2000s, there was a need for more realistic and accurate modelling of such events at NPPs as abnormal operation (events the frequency of which may exceed 10^{-2} event/ year). After analysis of the obtained results of a probabilistic safety analysis, such events include accidents involving the spillage of liquid radioactive materials (LRMs).

In general scope of investigation on the problem of radiological impact assessment for accidents with LRM leakage, the following tasks should be highlighted:

- identify possible scenarios of emergency processes on the example of foreign and domestic experience, actual results of probabilistic safety analysis for Ukrainian NPPs;
- determine the general conditions and characteristics of atmospheric releases, the features of mitigation strategies, and the reduction of adverse effects in such accidents;
- to develop a comprehensive mathematical model for the transport of radioactive material in emergency room;
- evaluate the levels of radioactive contamination at the site and the quantitative characteristics of the release into the environment on known scenario,
- integrate the developed model with the existing software tools for estimation of radiation doses of personnel and the population, environmental pollution;
- provide testing of the development on the basis of a number of demonstration calculations on representative accident scenarios and on the real situation.

Some tasks from this list are already investigated.

Worldwide there have been more than 30 significant accidents with spills of liquid radioactive materials at nuclear fuel cycle facilities for the last 60 years according to [21]. Among there is the accident at research reactors, nuclear power plants, nuclear complexes, uranium facilities, pilot plants and chemical plants, etc.

Many special computer codes and methods are currently developed to assess, with sufficient accuracy, parameters of releases for various accidents at nuclear fuel cycle facilities (e.g. MELCOR computer code manual, US NRC for NPP). But, according to study [21], these codes have some disadvantages and often require a large amount of input data, calculation time.

This papers is on development of the effective tool on source term calculation that allow to provide conservative results for such class of events as spill of LRMs in areas with forced ventilation.

3 Design Background

Atmospheric dispersion modelling provides a lot of requirements be taken into account in source term modeling for event associated with LRM. E.g., local scale model chains of DSS JRODOS allow to use several ways to fill in input parameters window. Most of them are used to bind to the initial activity (the number fission products) of the NPP reactor core. However, for cases where source term is undetermined by default user library, such as wildfire in the contaminated forest, unintentional melting sources of ionizing radiation, emissions of pharmaceuticals and etc., it is possible to enter initial parameters at some number of time intervals (e.g., in the versions of DSS JRODOS 2017, 2019—up to 128 time steps).

It is important to notice in most cases territory affected due to LRM spills are cordoned in near range scale. So scaling effects related to it (e.g. building wake effect shown in Fig. 1) should be considered in atmospheric dispersion modelling.

Moreover, selected time step of source term should be harmonized with both spatial resolution of modelling and time resolution of numerical weather data. Given

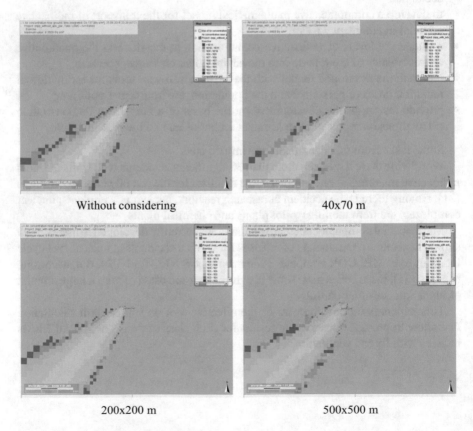

Without considering	40x70 m
200x200 m	500x500 m

Fig. 1 Building wake effect (4 cases with different sizes of building)

the resolution of numerical meteorological data for the model RIMPUFF, a minimum results time step that correlates with local sensitivity JRODOS atmospheric dispersion models and meteorological pre-processor is 10 min. Such step allows to provide an accurate calculation in near and middle range up to 21 h. In the case of input of stationary meteorological fields (manually) time step of the input data the emission source needs additional adjustments in accordance with the minimal time step of meteorological data.

Practice of sensitivity analysis of such primary ones atmospheric propagation results as the integral radionuclide concentration in the ground layer to the general JRODOS initial data package in [22], shows that the main problem in modelling is uncertainty in the source term. Such data include power, chronological parameters, speed, temperature, geometric height, radionuclide composition and physicochemical forms of emission over time. For import source data from the outside or export to a file, the JRODOS system uses xml-formats.

The main task of source term modelling is to obtain multiple arrays of parameters depending on the time variable with the help of individual analytical and/or numerical means. At the same time, a analysis of existing computer tools in [21] has shown to be a modern toolkit there are a number of significant drawbacks to modelling that do not allow it apply it to the quantitative assessment of discharges into the amosphere in case of LRM spills.

4 LRM as an Object of Study

LRM are liquid solutions, which include impurities of radioactive elements (possibly bound in high-molecular complexes). The isotopic composition of LRM is determined primarily by the source of radioactive impurities.

The main sources of LRM at nuclear power plants and nuclear complexes are as follows:

- primary coolant that is discharged for operational reasons;
- water that is used to back flush filters and ion exchangers;
- floor drains that collect water that has leaked from the active liquid systems and fluids from the decontamination of the plant and fuel flasks;
- leaks of secondary coolant;
- laundries and changing room showers;
- chemical laboratories.

At the same time, LRM can be located: both under containment of NPP units and beyond (for example, in an auxiliary building). At Ukrainian power units, temperatures of the LRM can reach 320 °C (under pressure), fluctuations in the range from 40 to 100 °C are possible in pipelines and tanks, depending on the ways of discharge of radioactive effluents [23, 24]. Accidents involving LRM spills are characterized by intense heat transfer due to the evaporation of the liquid—the formation of vapor-aerosol forms, which are subsequently localized on the materials of treatment or localization systems—for example, on drops of a sprinkler system or on gas-aerosol filters of ventilation equipment. In case of disability of localizing systems, significant emissions of radionuclides can be considered due to the leakage from the emergency rooms.

The isotopic composition and activity of the LRM at the NPP varies greatly. E.g., the primary coolant and the water of NPP spent fuel pool holding with light water reactors makes a collection of fragments of forced separation ^{235}U and ^{238}U, isotopes of corrosion metals, neutron activation products, etc. Within an auxiliary building, radioactive media can be maintained for a long time and include only long-lived radionuclides (^{60}Co, ^{134}Cs, ^{137}Cs, ^{54}Mn etc.) in the isotope composition. A similar case is observed in research reactors [25]. Heavy water, which is used as a moderator on liquid reactors, also has high activity due to the presence of tritium in it [26].

World nuclear complexes produce and process radioactive materials (e.g.NPP fuel, isotopic mixtures, industrial and medical sources). At nuclear complexes, as well as at nuclear power plants, reactor installations of different capacity are used. Nuclear complexes, chemical plants and research centers also work with the LRM. A distinctive feature of these enterprises in comparison with the NPP, in terms of the characteristics of the LRM used, is the large range of radioactive solutions involved in the technological processes of the enterprise.

At nuclear power plants and nuclear complexes, LRM are mainly aqueous solutions of decay products of nuclear fuel according to IAEA safety standards [27] and FSUE's materials [28]. Pilot plants and chemical plants may contain a full range of isotopes and solvents. LRM can be found on the site for processing the radioactive liquids.

Issue is a significant problem for radiotherapy medical hospitals. The problem is to analyze the emergency situations related to the special reservoir systems in the underground rooms of hospital (Decay Tank Systems containing ^{99m}Tc, ^{131}I, ^{18}F shown in Fig. 2).

Fig. 2 Resevoir of Decay Tank System (DTS) in Vienna SMZ-Ost Hospital (2 photos, October 2018)

5   Model Overview

The proposed approach is based on the theory of non-stationary heat and mass transfer in surface evaporation of liquid heated below the boiling temperature. The physical model includes: active liquid medium, steam-aerosol radioactive mixture (SARM), air of forced ventilation, airborne filters, and the floor of emergency area. The key aspects of the model are evaporation of liquid material, its removal with exhaust ventilation and partial trapping on airborne filters. It is considered that SARM is released to the environment after filters.

The model is developed to assess the radiation consequences in an accident with spill of LRM and describes the spread of radioactive material by two consecutive ways: releases within the emergency area and further transfer of SARM into the atmosphere. In this paper, we focus on the relationship of these pathways and study the process of evaporation as special one for accident with spill of LRM.

5.1   Basic Principles and Assumptions of the Model

The model of LRM evaporation describes the transport of radionuclides within the emergency area. It takes into account processes such as:

- evaporation of the radioactive material;
- drainage of LRM;
- cooling of LRM by evaporation and heat transfer from floor;
- entrainment of SARM via ventilation;
- cleaning of SARM to remove radioactive aerosols on the filtration facility;
- release of LRM to the atmosphere.
- Key assumptions in modeling the thermal evaporation processes:
- temperature of LRM is always higher than the temperature of involved air;
- condensation of SARM on surfaces of equipment and building designs is neglected;
- pressure in the airspace of the area is constant and equals atmospheric pressure;
- LRM's heat transfer coefficient relative to the floor does not depend on LRM temperature;
- convection flows within LRM are absent;
- concentration gradient of radionuclides in LRM is absent;
- heat transfer through the evaporation surface to the air of ventilation is neglected;
- velocity fields of aerodynamic flow and thermodynamic parameters of the air over the entire surface of evaporation are constant.

5.2   Description of LRM Evaporation

The process of evaporation that occurs in direct contact of the supplied air and liquid surface is complex. It combines the effects of heat and mass transfer. The mass flow of SARM released from the surface of the liquid is determined by the Dalton equation according to Nesterenko [29]:

$$dm_w = -\beta_{sw} \cdot (p_{sw} - p_m) \cdot Sdt \tag{1}$$

where m_w—mass of LRM, kg;

β_{sw}—mass transfer coefficient for normal atmospheric pressure, $\frac{kg}{Pa \cdot m^2 \cdot s}$;

p_{sw}, p_m—saturation pressures of LRM for temperature of liquid surface T_{sw} and temperature of involved air T_m (Fig. 1), Pa;

S—area of evaporation surface, m^2;

t—time, s.

Accordingly, the heat flow which is directed from the liquid to the supplied air is:

$$dQ_{ev} = -r_w \cdot \beta_{sw} \cdot (p_{sw} - p_m) \cdot Sdt \tag{2}$$

where r_w—heat of LRM evaporation, J/kg

In this case, if the temperature of LRM is always higher than the temperature of the supplied ventilation air in the boundary liquid layer, it gives rise to the temperature gradient, the nature of which depends on the intensity of heat and mass transfer.

The experimental data [30] show that the intensity of water transfer during evaporation depends on hygrothermal state of the incoming air flow (Fig. 3).

On the basis of experimental studies [30] the dependence of surface temperature of the evaporating liquid on hygrometric conditions and hydrodynamic process when the heat flow is directed from liquid surface to the environment was established. According to formulas (1) and (2), the parameters β_{sw} and p_{sw} are functions of temperature of the liquid surface T_{sw}, and value r_w is functions of temperature within the liquid T_w.

Fig. 3 The temperature layers near the surface of LRM

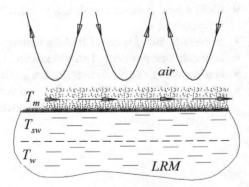

The model of evaporation also describes the heat transfer from liquid to the floor of emergency area. According to Isachenko [31], the heat transfer from the liquid to the floor is described by the expression:

$$dQ_f = k \cdot F \cdot (T_w - T_f)dt \tag{3}$$

where k—heat transfer coefficient for floor (this value can be found according to Isachenko [31]), $\frac{W}{m^2 \cdot K}$;
F—area of contact surface of LRM with the floor, m^3.
Then the heat balance for LRM can be written as:

$$dQ_w = -dQ_{ev} - dQ_f \tag{4}$$

At the same time:

$$dQ_w = c_p m_w dT_w \tag{5}$$

where c_p—isobaric heat capacity of LRM (value c_p is function of temperature T_w), $\frac{J}{kg \cdot K}$.
In view of formulas (2), (3) and (5), heat balance for LRM (4) can be written as:

$$c_p m_w dT_w = -r_w \beta_{sw}(p_{sw} - p_m)Sdt - kF(T_w - T_f)dt \tag{6}$$

Functions $\beta_{sw}(T_{sw})$, $p_{sw}(T_{sw})$, $p_m(T_m)$, $r_w(T_w)$ and $c_p(T_w)$ are polynomials. They are compiled according to Volkov [32] and Rivkin & Aleksandrov [33].

In model, the performance of forced ventilation unit is determined by the flowrate of involved air. After the exhaust ventilation pipe, SARM partially settles on filters. This phenomenon accounts for the coefficient of filtration. It determines the relative amount of SARM that is deposited on the filter material. SARM further passes through the ventilation stack into the atmosphere.

Another method to confine LRM is to drain the spilled liquid by drainage pumps or by gravity. This process is characterized by the flowrate of LRM through the drainage channel.

The partial removal of radioactive substances from the liquid by evaporation depends on physicochemical properties of radioactive impurities and the solvent.

5.3 Modeling of the LRM Evaporation

To solve the problem of unsteady LRM evaporation, using Eqs. (1)–(6), four balance differential Eqs. (7) were written to relate the main parameters of LRM and air space of area over time:

$$\begin{cases} \dfrac{dm_w}{dt} = -\beta_{sw}(p_{sw} - p_m)S - G_d \\[2mm] \dfrac{dm_a}{dt} = \beta_{sw}(p_{sw} - p_m)S - G_V \cdot \dfrac{m_a}{V} \\[2mm] \dfrac{dm_q}{dt} = G_V \dfrac{m_a}{V}(1 - \psi) \\[2mm] \dfrac{dT_w}{dt} = -\dfrac{r_w\beta_{sw}(p_{sw} - p_m)S + kF\left(T_w - T_f\right)}{c_p m_w} \end{cases} \tag{7}$$

Where m_a—current mass of SARM in air of the area, kg;

G_d—flowrate of LRM through the drainage channel (it also includes the volume of LRM leakage from the area), kg/s;

V—air volume in the area, m^3;

G_V—flowrate of involved air of forced ventilation (this parameter includes SARM leakage through the gaps or clearances in walls of the emergency area) m^3/s;

ψ—coefficient of filtration (efficiency of filtration);

m_q—mass of released SARM into the atmosphere, kg

This system of nonlinear differential equations includes polynomial functions. Using the Mathcad sphere for solving the system of Eqs. (7) provides the desired functions in matrix form (the values of the functions at particular moments of accident).

Average activity concentration of the radionuclide in the area air A_{air} (Bq/m^3) is given by the formula

$$A_{air} = \frac{A_w}{V} H \cdot m_a \tag{8}$$

where A_w—concentration of radionuclide in LRM, Bq/kg;

H—fraction of carried away solute with solvent vapors during evaporation.

The ultimate objective of the model is to determine the dynamics of LRM evaporation, SARM activity in the air space and the integral release of radioactive substances into the atmosphere. The mass fraction of a radionuclide in the release relative to its original content in radioactive liquid is commonly used in practice:

$$q = \frac{A_w}{m_0} H \cdot m_q \cdot 100\% \tag{9}$$

where m_0—initial mass of LRM, kg.

This value is used as an input parameter for the assessment of doses to the public from atmospheric release.

6 Analysis of the Influence of Initial Accident Conditions on the Results of Assessment

During the study, radiation consequences were assessed with the aim of briefly analysis of some results obtained for the hypothetical accident with spills of LRM in the storage area of liquid radioactive waste with a forced ventilation system, according to the input data specified in Table 1.

SARM released into the atmosphere through the ventilation stack (100 m height) in the following weather conditions:

wind speed of 1 m/s;

class of atmospheric stability "B" (the most conservative class for release height of 100 m and receptor at 2.5 km as the typical radius of sanitary protection zones around Ukrainian NPPs);

roughness of the underlying surface soil—10 cm.

To analyze the influence of the initial accident conditions on the final results of the assessment, a number of input parameters were chosen: flowrate (G_V) and temperature (T_m) of supplied ventilation air, LRM drainage flowrate (G_d). Unlike other parameters, these values can be adjusted during an accident. These data are represented as a series of values in Table 2.

Some results of the assessment of radiation consequences are shown in Table 3 (for the mix $^{137}Cs + {}^{60}Co$) and Figs. 4, 5 and 6 (only for ^{137}Cs).

Curves of Figs. 4, 5 and 6 for ^{137}Cs are similar to ^{60}Co, cause, according to formula (8), the values of activity concentration A_{air} of both radionuclide's are linearly dependent. According to formula (9), it is the same for mass fraction q. Therefore, for the analysis we showed the curves only for ^{137}Cs.

According to the assessment results (Fig. 4a), with higher flow rate, the activity concentration in the air of emergency area reduces and the release of radionuclides into the atmosphere (Fig. 4b) slightly increases with further growth of the flow rate.

Table 1 Input data of assessment

#	Parameter	Value
1	Initial mass of LRM m_0, kg	$2.0 \cdot 10^5$
2	Initial temperature of LRM $T_{w0}, {}^oC$	98
3	Area of evaporation S, m^2	200
4	Area of contact surface of LRM with the floor F, m^2	210
5	Average temperature of the floor T_f, $°C$	20
6	Air volume in the area V, m^3	1000
7	Efficiency of filtration ψ	0
8	Heat transfer coefficient for the floor k, $\frac{W}{m^2 \cdot K}$	15.4
9	Average activity concentration of radionuclide in LRM A_w, Bq/kg: ^{137}Cs ^{60}Co	$3.7 \cdot 10^8$ $4.7 \cdot 10^7$

Table 2 Input parameters of operating modes of ventilation and drainage systems

#	$G_V, m^3/s$	$T_m, {}^0C$	$G_d, kg/s$
1	0.001	10	0
2	0.75		
3	1.5		
4	2.25		
5	0.75	5	0
6		25	
7		45	
8	0.75	10	0
9			50
10			100
11			150

Table 3 Values of the maximum effective dose to the human body on distance 2.5 km from release point during 14 days depending on the operating modes of ventilation and drainage systems

#	Effective dose $D_{ef}, \mu Sv$
1	1.94
2	14.90
3	15.53
4	16.15
5	16.50
6	15.53
7	12.53
8	14.83
9	3.13
10	1.88
11	1.39

Therefore, there is the minimum flow rate that may provide sufficient reduction in SARM activity in the air of area. In this accident, atmospheric release is eliminated only if the ventilation system is inactive and the emergency area is completely confined.

The results presented in Fig. 5 indicate that heating of supply ventilation air (to the temperature of LRM surface T_{sw}) helps to suppress the evaporation process. The activity concentration of SARM in the air of emergency area and integral SARM release into the atmosphere decrease with increasing temperature of supply air.

The calculated data show (Fig. 6) that drainage of LRM is the most effective way to confine and prevent the release of radioactive substances into the air of emergency area and the environment. In the absence of drainage (curve 1, Fig. 6b), the integral release of SARM is rapidly growing even in 2 h from the beginning of the accident.

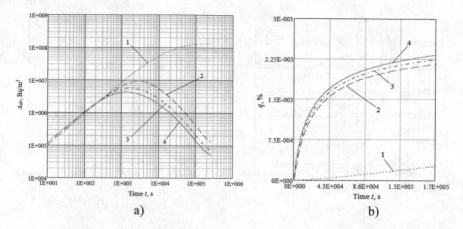

Fig. 4 Changes in average activity concentration in the air of emergency area (**a**) and mass fraction of ^{137}Cs released to the environment (**b**) for the following ventilation flow rates: $G_V = 0.001\ m^3/s$ (curve 1); $G_V = 0.75\ m^3/s$ (curve 2); $G_V = 1.5\ m^3/s$ (curve 3); $G_V = 2.25\ m^3/s$ (curve 4)

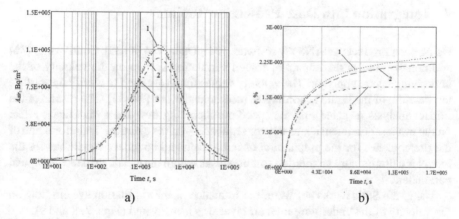

Fig. 5 Changes in average activity concentration in the air of emergency area (**a**) and mass fraction of ^{137}Cs released to the environment (**b**) for the following supply air temperature: $T_m = 5\ °C$ (curve 1), $T_m = 25\ °C$ (curve 2), $T_m = 45\ °C$ (curve 3)

It is interesting to note that curves 2–4 in Fig. 6a are not smooth. Each of them has the point of "breaking". At these times, the confinement of LRM and the evaporation process are terminated. Clearing of SARM remaining in the air of area begins.

In all cases, the effective dose to the public and personnel from a particular radionuclide is directly proportional to the size of release and activity concentration in the emergency area, respectively. Assessment of effective doses to the public has showed that this accident cannot cause significant radiation consequences (Table 3). High flow rate and low temperature of supplied ventilation air contribute to increase effective doses to the public.

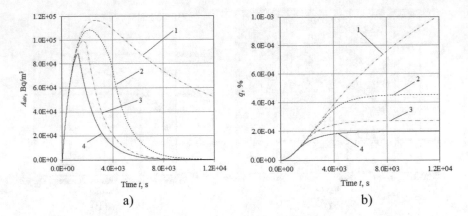

Fig. 6 Changes in average activity concentration in the air of emergency area (**a**) and mass fraction of ^{137}Cs released to the environment (**b**) for the following drainage flow rates: $G_d = 0$ kg/s (curve 1), $G_d = 50$ kg/s (curve 2), $G_d = 100$ kg/s (curve 3), $G_d = 150$ kg/s (curve 4)

7 Integration into Dose Projection Toolkit

Using such products as ANSYS software [34], OpenFOAM [35], SolidWorks [36] additionally studies the impact of forced ventilation system on the velocity of the underlying layer supply air. The velocity values found are used to clarify the surface temperature of the liqiud according to recommendations [29–32]. The results of the impact analysis revealed that the speed the incoming flow has a significant effect on the evaporation intensity, as the consequence—on the ejection. Therefore, one of the prerequisites for the preparation of initial data and boundary conditions for the model is a preliminary ventilation solution tasks within the established technological parameters.

Using the SolidWorks software, the influence of the ventilation system flow on the velocity of the underlying inlet air layer was investigated (Figs. 7, 8 and 9).

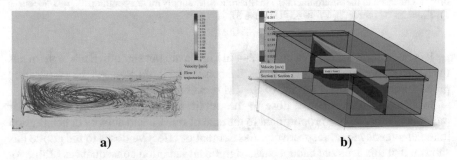

Fig. 7 Air flow fields SolidWorks simulation ($G_V = 0{,}75$ m^3/s): flow trajectories (**a**), velocity fields (**b**)

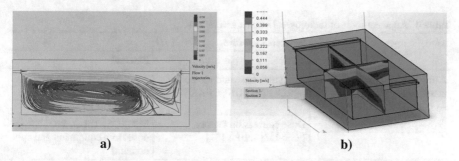

Fig. 8 Air flow fields SolidWorks simulation ($G_V = 1,5 \text{ m}^3/\text{s}$): flow trajectories (**a**), velocity fields (**b**)

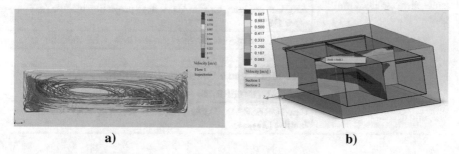

Fig. 9 Air flow fields SolidWorks simulation ($G_V = 2,25 \text{ m}^3/\text{s}$): flow trajectories (a), velocity fields (b)

Based on the results of the mathematical model with a view to further determination of radiological impact on the wokers, public and the environment the environment can be used analytical methods and software tools (Table 4).

8 Testing

Developed model was succesfully validated and verified on real case happened in 2017 on one of Pakistan NPPs units. As it was announced on IAEA web-site https://www.iaea.org/, on August 04, Karachi NPP unit 1 was in shutdown state when bonnet gasket of moderator drain tank outlet valve was replaced as per preventive maintenance program. After the maintenance work, moderator drain tank was slowly filled with heavy water on August 04, 2017 while its outlet valve was kept close as per procedure. After about 2 h of moderator drain tank filling, drop-wise leak of heavy water started from its outlet valve bonnet gasket that gradually kept on increasing. To save heavy water from leaking, tank outlet valve was manually opened and heavy water was transferred to dump space. Boiler room was on purging with fresh air

Table 4 Adjacent tools of radiological impact assessment

Model outputs	Workers exposure	
	Internal	External
Array of air concentrations in emergency room $A_V(t)$	Analytical base and methods (NRC, ICRP, UNSCEAR [37]);	Analytic base (NRC, EPA, ICRP) + dose from skin exposure (α and β);
	Dose conversion factors FGR-11/13, EPA [37]	MICROSHIELD, ISOCSR (+ dose from equipment and spill domain)
	Public exposure and environmental contamination	
Array of time-integrated activity released into the atmosphere (source term) $Q(t)$	RODOS: ADM RIMPUFF with 10-min. time step + FDMT [38];	
	ARGOS: complex terrain ADM, dose projection module (any other DSS);	
	sophisticated models for short range (CFD-, LES-modeling);	
	samplified gaussian models;	
	NRC MACCS code (probabilistic tool) [39];	
	RASCAL (INTERRAS) [40], HOTSPOT [41];	
	GENII, RESRAD, PAVAN, ARCON 96, XOQDOQ (RAMP family) [37]; etc.	

before this leak to carry out maintenance jobs. Purging was terminated and spilled heavy water was collected.

In response to the requirement of the regulatory body, Karachi NPP unit 1 performed Root Cause Analysis (RCA) of this event and submitted the report on October 02, 2017. According to this report, about 1589 kg spilled heavy water of moderator system was collected from reactor building floor. Among the workers involved in valve isolation and heavy water collection, four individuals exceeded the annual regulatory dose limit of 20 mSv/year.

Dose reconstruction was provided using the developed model and was compared with the actual value of the dose recieved by the liquidators over time of works (Table 5).

Analysis of obtained testing resultes shows that

Table 5 Modeling and measurement results

Person	Effective dose (modelling results), mSv	Effective dose received by workers based on measurements, mSv	Multiple for pessimistic results
#1	6.9...43.0	20.8(+−30%)	2.07
#2	9.3...57.6	24.2(+−30%)	2.38
#3	11.4...71.0	30.9(+−30%)	2.30
#4	12.6...78.7	36.2(+−30%)	2.17

- all modelling results is within the estimated effective dose range, which is acceptable;
- the pessimistic estimate exceeded the actual measurement data ~ 2.0–2.3 times, which is acceptable;
- the developed model can be used as an effective tool for dose projection.

9 Conclusions

For the first time a mathematical model for both analysis and prediction of the source term for LRM spills taking into account technological premises and adjacent environmental has been proposed and developed. This model, unlike other models, takes into account the parameters of the radioactive liquid composition and the design conditions of their storage.

Possible ways of software integration of the developed mathematical model into dose orojection toolkit, taking into account the requirements of the adjacent radiation impact assessment tool and the specificity of liquid radioactive accident spill, is proposed. It solve the problem of complex radological estimation for events including LRM.

On the example of a series of emergency scenarios for disturbances of normal operation at the NPP, the sensitivity of the simulation results to the change of initial parameters was investigated, which allowed to test the developed model and to rank the considered physical processes by their relevance.

The developed model can be applied as useful tool of source term preparation in the case of LRM spill in advance asa well as in real time with a reasonable degree of conservatism. It allows to provide an adequate assessment of the radiation consequences for the set of initial conditions of accidents with spills of LRM in areas with forced ventilation.

References

1. Popov, O., Iatsyshyn, A., Kovach, V., Artemchuk, V., Taraduda, D., Sobyna, V., Sokolov, D., Dement, M., Yatsyshyn, T., Matvieieva, I.: Analysis of possible causes of npp emergencies to minimize risk of their occurrence. Nucl. Radiat. Saf. **81**(1), 75–80 (2019). https://doi.org/10.32918/nrs.2019.1(81).13
2. Popov, O., Iatsyshyn, A., Kovach, V., Artemchuk, V., Taraduda, D., Sobyna, V., Sokolov, D., Dement, M., Yatsyshyn, T.: Conceptual approaches for development of informational and analytical expert system for assessing the NPP impact on the environment. Nuc. Radiat. Saf. **79**(3), 56–65 (2018). https://doi.org/10.32918/nrs.2018.3(79).09
3. Norms of Radiation Safety (NRBU-97). http://www.insc.gov.ua/docs/nrbu97.pdf
4. IAEA-TECDOC-1200 Applications of probabilistic safety assessment (PSA) for nuclear power plants, IAEA, Vienna (2006)
5. IAEA-TECDOC-1511 Determining the quality of probabilistic safety assessment (PSA) for applications in nuclear power plants, IAEA, Vienna (2006)

6. Energy Strategy of Ukraine for the period up to 2035 "Security, Energy Efficiency, Competitiveness". https://zakon.rada.gov.ua/laws/file/text/58/f469391n10.pdf

7. Implementation of the Comprehensive (Integrated) Program of Safety Improvement of Power Units of Nuclear Power Plants. https://zakon.rada.gov.ua/laws/show/1270-2011-%D0%BF?lang=en

8. Bogorad, V., Bielov, Y., Kyrylenko, Y., Lytvynska, T., Poludnenko, V., Slepchenko, O.: Forecast of the consequences of a fire in the chernobyl exclusion zone: A combination of the hardware of the mobile laboratory RanidSONNI and computer technologies DSS RODOS. Nucl. Radiat. Saf. **79**(3), 10–15 (2018). https://doi.org/10.32918/NRS.2018.3(79).02

9. Popov, O., Iatsyshyn, A., Kovach, V., Artemchuk, V., Taraduda, D., Sobyna, V., Sokolov, D., Dement, M., Hurkovskyi, V., Nikolaiev, K., Yatsyshyn, T., Dimitriieva, D.: Physical features of pollutants spread in the air during the emergency at NPPs. Nucl. Radiat. Saf. **84**(4), 88–98 (2019). https://doi.org/10.32918/nrs.2019.4(84).11

10. Pereira, M.N.A., Schirru, R., Gomes, K.J., Cunha, J.L.: Development of a mobile dose prediction system based on artificial neural networks for NPP emergencies with radioactive material releases. Ann. Nucl. Energy **105**, 219–225 (2017). https://doi.org/10.1016/j.anucene.2017.03.017

11. Raja Shekhar, S.S., Venkata Srinivas, C., Rakesh, P.T., Deepu, R., Prasada Rao, P.V.V., Baskaran, R., Venkatraman, B.: Online Nuclear Emergency Response System (ONERS) for consequence assessment and decision support in the early phase of nuclear accidents—simulations for postulated events and methodology validation. Prog. Nucl. Energy **119** (2020). https://doi.org/10.1016/j.pnucene.2019.103177

12. Kovalets, I.V., Robertson, L., Persson, C., Didkivska, S.N., Ievdin, I.A., Trybushnyi, D.: Calculation of the far range atmospheric transport of radionuclides after the Fukushima accident with the atmospheric dispersion model MATCH of the JRODOS system. Int. J. Env. Poll. **54**(2–4), 101–109 (2014). https://doi.org/10.1504/IJEP.2014.065110

13. Zabirov, A.R., Smirnova, A.A., Feofilaktova, Y.M., Shevchenko, S.A., Yashnikov, D.A.: Russian experimental database for validation of computer codes used for safety analysis of nuclear facilities. Prog. Nucl. Energy **118** (2020). https://doi.org/10.1016/j.pnucene.2019.103061

14. Hu, L., et al.: SuperMC cloud for nuclear design and safety evaluation. Anna. Nucl. Energy **134**, 424–431 (2019). https://doi.org/10.1016/j.anucene.2019.07.019

15. Mohammed Saeed, I.M., Mohammed Saleh, M.A., Hashim, S., bin Ramli, A.T., Al-Shatri, S.H.H.: Atmospheric dispersion modeling and radiological safety assessment for expected operation of Baiji nuclear power plant potential site. Ann. Nucl. Energy **127**, 156–164 (2019). https://doi.org/10.1016/j.anucene.2018.11.045

16. Tabadar, Z., Ansarifar, G.R., Pirouzmand, A.: Thermal-hydraulic modeling for deterministic safety analysis of portable equipment application in the VVER-1000 nuclear reactor during loss of ultimate heat sink accident using RELAP5/MOD3.2 code. Ann. Nucl. Energy **127** 53–67 (2019). https://doi.org/10.1016/j.anucene.2018.11.046

17. Zaporozhets, A.: Analysis of control system of fuel combustion in boilers with oxygen sensor. Peri. Polytechn. Mechan. Eng. **63**(4), 241–248 (2019). https://doi.org/10.3311/PPme.12572

18. Zaporozhets, A.: Development of sowware for fuel combustion control system based on frequency regulator. In: CEUR Workshop Proceedings, vol. 2387, pp. 223–230. Online: http://ceur-ws.org/Vol-2387/20190223.pdf

19. Popov, O.O., Iatsyshyn, A.V., Kovach, V.O., Artemchuk, V.O., Kameneva, I.P., Taraduda, D.V., Sobyna, V.O., Sokolov, D.L., Dement, M.O., Yatsyshyn, T.M.: Risk assessment for the population of Kyiv, Ukraine as a result of atmospheric air pollution. J. Health Poll. **10**(25), 200303 (2020). https://doi.org/10.5696/2156-9614-10.25.200303

20. Norms of Radiation Safety of Ukraine. Supplement: Radiation Protection from Potential Ionising Radiation Sources (NRBU-97/D-2000). https://zakon.rada.gov.ua/rada/show/v01164 88-00

21. Kyrylenko, Y., Kameneva, I.: The problem of radiation impact evaluation for accidents with spills of liquid radioactive materials. Model. Inf. Technolo. **82**, 52–64 (2018)

22. Kyrylenko, Y., Kameneva, I.: Computer tools for a radiation accidents and abnormal operations consequences modelling at NPP. Model. Inf. Technol. **84**, 79–87 (2018)
23. Kyrylenko, Y., Kameneva, I.: Input data preparation for radiological impact modelling problems in case of accidents involving of liquid radioactive materials spills. Collection of works for Conference « Simulation-2018 » , pp. 162–165 (2018)
24. The radiochemistry of nuclear power plants with light water reactors. Karl-Heinz Neeb, Handbook—Berlin: New York (1997)
25. Analysis Report. Analysis of design basis accidents. Rivne NPP (2017)
26. Security Analysis Report: WWR-M Research Reactor. Institute for Nuclear Research of the NAS of Ukraine, Kyiv (2016)
27. Safe Handling of Tritium Review of Data and Experience. Technical Reports Series No. 324, IAEA (1991)
28. IAEA-TECDOC-1267 Procedures for conducting probabilistic safety assessment for non-reactor nuclear facilities, IAEA, Vienna (2002)
29. Materials on environmental impact assessment of the proposed activity on the operation of a nuclear facility, a complex of nuclear materials intended for radiochemical reprocessing of spent nuclear fuel. Federal State Unitary Enterprise "Production Association "Mayak" FSUE" PA "Mayak" (2012)
30. Nesterenko, A.: Fundamentals of thermodynamic calculations ventilation and air conditioning. Vysshaia Shkola, Moscow (1971)
31. Nesterenko, A. Experimental study of heat and mass transfer during evaporation of the liquid with a free surface. *Material science*. Technical Physics. Academy of Sciences of the USSR 1954. Vol. 24, No. 4
32. Isachenko, V., Osipova, V., Sukomel, A.: Heat Transfer. Energiya, Moscow (1975)
33. Volkov, O.: Designing of Industrial Building Ventilation. Vyscha Shkola, Kharkov (1989)
34. Rivkin, S., Aleksandrov, A.: Thermodynamic Properties of Water and Steam. Directory. Energiya, Moscow (1984)
35. ANSYS FLUENT 12.0 User's Guide. ANSYS, Inc. is certified to ISO 9001: pp. 2009—2070 (2008)
36. Greenshields, C.: OpenFOAM User Guide version 6. The OpenFOAM Foundation (2018)
37. RASCAL 4.3 User's Guide / Ramsdell Environmental Consulting, LLC (2013)
38. SolidWorks Flow Simulation. https://hawkridgesys.com/solidworks/
39. Sandia National Laboratories. https://www.sandia.gov
40. HotSpot Health Physics Codes Version 3.0 User's Guide/National Atmospheric Release Advisory Center, LNLL (2014)
41. Raskob, W., Landman, C., Trybushnyi, D.: Functions of decision support systems (JRodos as an example): overview and new features and products. Radioprotection **51**, number HS1 (2016). https://doi.org/10.1051/radiopro/2016015
42. WinMACCS, a MACCS2 Interface for Calculating Health and Economic Consequences from Accidental Release of Radioactive Materials into the Atmosphere MACCS User's Guide/U.S. Nuclear Regulatory Commission (2007)

Printed in the United States
by Baker & Taylor Publisher Services